Statistics Be the Headlines

How do you learn about what's going on in the world? Did a news headline grab your attention? Did a news story report on recent research? What do you need to know to be a critical consumer of the news you read? If you are looking to start developing your data self-defense and critical news consumption skills, this book is for you! It reflects a long-term collaboration between a statistician and a journalist to shed light on the statistics behind the stories and the stories behind the statistics. The only prerequisite for enjoying this book is an interest in developing the skills and insights for better understanding news stories that incorporate quantitative information.

Chapters in **Statistics Behind the Headlines** kick off with a news story headline and a summary of the story itself. The meat of each chapter consists of an exploration of the statistical and journalism concepts needed to understand the data analyzed and reported in the story. The chapters are organized around these sections:

- What ideas will you encounter in this chapter?

- What is claimed? Is it appropriate?

- Who is claiming this?

- Why is it claimed? What makes this a story worth telling?

- Is this a good measure of impact?

- How is the claim supported?

  - What evidence is reported?

  - What is the quality/strength of the evidence?

- Does the claim seem reasonable?

- How does this claim fit with what is already known?

- How much does this matter?

- Considering the coverage

Chapters close with connections to the Stats + Stories podcast.

A. John Bailer was University Distinguished Professor and Chair in the Department of Statistics at Miami University and an affiliate member of the Departments of Biology, Media, Journalism and Film and Sociology and Gerontology. His interests include promoting quantitative literacy and enhancing connections between statistics and journalism which resulted in the award-winning Stats + Stories podcast that he started with journalism colleagues in 2013.

Rosemary Pennington is Associate Professor in the Department of Media, Journalism and Film at Miami University. Her research examines the ways that marginalized groups are represented in media as well as how members of such groups may use media to challenge those representations. Pennington was a public broadcasting journalist working in Athens, Ohio, and Birmingham, Alabama.

ASA-CRC Series on
Statistical Reasoning in Science and Society

Series Editors

Nicholas Fisher, University of Sydney, Australia
Nicholas Horton, Amherst College, MA, USA
Regina Nuzzo, Gallaudet University, Washington, DC, USA
David J Spiegelhalter, University of Cambridge, UK

Published Titles

Data Visualization: Charts, Maps and Interactive Graphics
Robert Grant

Improving Your NCAA® Bracket with Statistics
Tom Adams

Statistics and Health Care Fraud: How to Save Billions
Tahir Ekin

Measuring Crime: Behind the Statistics
Sharon Lohr

Measuring Society
Chaitra H. Nagaraja

Monitoring the Health of Populations by Tracking Disease Outbreaks
Steven E. Fricker and Ronald D. Fricker, Jr.

Debunking Seven Terrorism Myths Using Statistics
Andre Python

Achieving Product Reliability: A Key to Business Success
Necip Doganaksoy, William Q. Meeker, and Gerald J. Hahn

Protecting Your Privacy in a Data-Driven World
Claire McKay Bowen

Backseat Driver: The Role of Data in Great Car Safety Debates
Norma F. Hubele

Statistics Behind the Headlines
A. John Bailer and Rosemary Pennington

For more information about this series, please visit: https://www.crcpress.com/go/asacrc

Statistics Behind the Headlines

A. John Bailer
Rosemary Pennington

CRC Press
Taylor & Francis Group
Boca Raton London New York

CRC Press is an imprint of the
Taylor & Francis Group, an **informa** business

A CHAPMAN & HALL BOOK

First edition published 2023
by CRC Press
6000 Broken Sound Parkway NW, Suite 300, Boca Raton, FL 33487-2742

and by CRC Press
4 Park Square, Milton Park, Abingdon, Oxon, OX14 4RN

CRC Press is an imprint of Taylor & Francis Group, LLC

Library of Congress Cataloging-in-Publication Data

Names: Bailer, A. John, author. | Pennington, Rosemary, author.
Title: Statistics behind the headlines / A. John Bailer, Rosemary Pennington.
Description: First edition published 2023 | Boca Raton : CRC Press, 2022. | Series: ASA-CRC series on statistical reasoning in science and society | Includes bibliographical references and index.
Identifiers: LCCN 2022016186 (print) | LCCN 2022016187 (ebook) | ISBN 9780367902537 (hardback) | ISBN 9780367902520 (paperback) | ISBN 9781003023401 (ebook)
Subjects: LCSH: Journalism--Mathematics. | Journalism--Statistics. | Statistics--Popular works.
Classification: LCC PN4784.M37 B35 2022 (print) | LCC PN4784.M37 (ebook) | DDC 070.4/495195--dc23/eng/20220414
LC record available at https://lccn.loc.gov/2022016186
LC ebook record available at https://lccn.loc.gov/2022016187

ISBN: 9780367902537 (hbk)
ISBN: 9780367902520 (pbk)
ISBN: 9781003023401 (ebk)

DOI: 10.1201/9781003023401

Typeset in Minion
by Deanta Global Publishing Services, Chennai, India

Contents

Preface, xv

Acknowledgments, xxiii

CHAPTER 1 ▪ A Field Guide to Reading
the *Statistics behind the Headlines*　　1

JOURNALISTS AND STATISTICIANS SHARE SIMILAR GOALS	2
STRUCTURE OF EACH CHAPTER	3
STATISTICAL CONCEPTS TO BE EXPLORED	4
WHAT IS IT? HOW MUCH IS THERE?	4
DATA GENERATION/PRODUCING DATA	5
DESCRIBING DATA	5
DRAWING CONCLUSIONS FROM DATA	6
IF I DO THIS, THEN THAT WILL HAPPEN	7
JOURNALISM 101	7

CHAPTER 2 ▪ Predicting Global Population
Growth and Framing How You Report It　　11

STORY SUMMARY	12
WHAT IDEAS WILL YOU ENCOUNTER IN THIS CHAPTER?	13

WHAT IS CLAIMED? IS IT APPROPRIATE? 13

WHO IS CLAIMING THIS? 14

WHY IS IT CLAIMED? WHAT MAKES THIS
A STORY WORTH TELLING? 15

IS THIS A GOOD MEASURE OF IMPACT? 15

HOW IS THE CLAIM SUPPORTED? 15

   What Evidence Is Reported? 17

   What Is the Quality/Strength of the Evidence? 18

DOES THE CLAIM SEEM REASONABLE? 18

HOW DOES THIS CLAIM FIT WITH WHAT IS
ALREADY KNOWN? 19

HOW MUCH DOES THIS MATTER? 19

   Comparison of Population Perspective versus
   Individual Perspective? 20

   Will I Change My Behavior as a Consequence of This? 20

CONSIDERING THE COVERAGE 21

REVIEW 24

STATS + STORIES PODCASTS 24

REFERENCES – WORLD POPULATION PROJECTION 25

NOTES 25

CHAPTER 3 ■ Social Media and Mental Health 27

STORY SUMMARY 28

WHAT IDEAS WILL YOU ENCOUNTER IN THIS
CHAPTER? 29

WHAT IS CLAIMED? IS IT APPROPRIATE? 29

WHO IS CLAIMING THIS? 30

WHY IS IT CLAIMED? 31

IS THIS A GOOD MEASURE OF IMPACT? 31

   Variables 31

   Odds and Odds Ratios 34

HOW IS THE CLAIM SUPPORTED?   35

What Evidence Is Reported?   38

What Is the Quality/Strength of the Evidence?   43

IS A 2X INCREASE IN ODDS OF PROBLEMS
A CAUSE FOR CONCERN?   45

WHAT ARE THE BASELINE RATES OF THESE
MENTAL HEALTH PROBLEMS?   46

IS THE CLAIM REASONABLE IN ITSELF?
DOES PRIOR BELIEF IMPACT MY BELIEF?
CONFIRMATION BIAS?   46

HOW DOES THIS CLAIM FIT WITH WHAT IS
ALREADY KNOWN?   46

HOW MUCH DOES THIS MATTER TO ME?   47

DOES A STUDY OF U.S. YOUNG TEENS TRANSLATE
TO OLDER TEENS OR TO OTHER COUNTRIES?   48

CONSIDERING THE COVERAGE   49

REVIEW   52

STATS + STORIES PODCASTS   52

NOTES   53

CHAPTER 4 ▪ Speedy Sneakers: Technological
Boosterism or Sound Science?   55

STORY SUMMARY   56

WHAT IDEAS WILL YOU ENCOUNTER IN THIS
CHAPTER?   57

WHAT IS CLAIMED? IS IT APPROPRIATE?   57

WHO IS CLAIMING THIS?   57

WHY IS IT CLAIMED?   58

IS THIS A GOOD MEASURE OF IMPACT?   60

HOW IS THE CLAIM SUPPORTED?   60

What Evidence Is Reported?   60

What Is the Quality/Strength of the Evidence?   63

IS THE CLAIM REASONABLE IN ITSELF?
DOES PRIOR BELIEF IMPACT MY BELIEF?
CONFIRMATION BIAS?   66

HOW DOES THIS CLAIM FIT WITH WHAT IS
ALREADY KNOWN?   66

HOW MUCH DOES THIS MATTER TO ME?   67

CONSIDERING THE COVERAGE   68

REVIEW   71

TO LEARN MORE   72

A BONUS STORY   72

STATS + STORIES PODCASTS   73

CHAPTER 5 ■ Investigating Series Binge-Watching   75

STORY SUMMARY   77

WHAT IDEAS WILL YOU ENCOUNTER IN THIS
CHAPTER?   77

WHAT IS CLAIMED? IS IT APPROPRIATE?   77

WHO IS CLAIMING THIS?   78

WHY IS IT CLAIMED?   78

IS THIS A GOOD MEASURE OF IMPACT?   79

HOW IS THE CLAIM SUPPORTED?   79

What Evidence Is Reported?   80

HOW MUCH TELEVISION DO YOU WATCH?
GOVERNMENT SURVEY SAYS …   80

ARE YOU A BINGE-WATCHER? INDUSTRY
REPORT SAYS …   82

IS WATCHING LOTS OF TV IS GOOD, BAD OR
BOTH FOR YOU? EXPERTS SAY …   84

BINGING AND STRESS? SCIENTIFIC
PRESENTATION SAYS …   84

What Is the Quality/Strength of the Evidence?   86

IS THE CLAIM REASONABLE IN ITSELF?
DOES PRIOR BELIEF IMPACT MY BELIEF?
CONFIRMATION BIAS?   86

HOW DOES THIS CLAIM FIT WITH WHAT IS
ALREADY KNOWN?   87

HOW MUCH DOES THIS MATTER TO ME?   87

CONSIDERING THE COVERAGE   87

REVIEW   89

STATS + STORIES PODCASTS   90

CHAPTER 6 ■ Tracking the Spread of "False News"   91

STORY SUMMARY   92

WHAT IDEAS WILL YOU ENCOUNTER IN THIS
CHAPTER?   93

WHAT IS CLAIMED? AND IS IT APPROPRIATE?   93

WHO IS CLAIMING THIS?   94

WHY IS IT CLAIMED?   94

IS THIS A GOOD MEASURE OF IMPACT?   95

HOW IS THE CLAIM SUPPORTED?   97

What Evidence Is Reported?   97

What Is the Quality/Strength of the Evidence?   98

IS THE CLAIM REASONABLE IN ITSELF?
DOES PRIOR BELIEF IMPACT MY BELIEF?
CONFIRMATION BIAS?   100

HOW DOES THIS CLAIM FIT WITH WHAT IS
ALREADY KNOWN?   100

HOW MUCH DOES THIS MATTER TO ME?   101

CONSIDERING THE COVERAGE   101

REVIEW   105

STATS + STORIES PODCASTS   105

NOTE   106

CHAPTER 7 ■ Modeling What It Means to
"Flatten the Curve"   107

  STORY SUMMARY   108

  WHAT IDEAS WILL YOU ENCOUNTER IN THIS
  CHAPTER?   109

  WHAT IS CLAIMED? IS IT APPROPRIATE?   109

  WHO IS CLAIMING THIS?   110

  WHY IS IT CLAIMED?   110

  IS THIS A GOOD MEASURE OF IMPACT?   111

  HOW IS THE CLAIM SUPPORTED?   111

    What Evidence Is Reported?   112

    What Is the Quality/Strength of the Evidence?   113

  IS THE CLAIM REASONABLE IN ITSELF?
  DOES PRIOR BELIEF IMPACT MY BELIEF?
  CONFIRMATION BIAS?   115

  HOW DOES THIS CLAIM FIT WITH WHAT IS
  ALREADY KNOWN?   115

  HOW MUCH DOES THIS MATTER TO ME?   116

  CONSIDERING THE COVERAGE   116

  REVIEW AND RECAP   119

  COVID CODA   119

  STATS + STORIES PODCASTS   120

CHAPTER 8 ■ One Governor, Two Outcomes
and Three COVID Tests   123

  STORY SUMMARY   124

  WHAT IDEAS WILL YOU ENCOUNTER IN THIS
  CHAPTER?   125

  WHAT IS CLAIMED? IS IT APPROPRIATE?   125

  WHO IS CLAIMING THIS?   126

WHY IS IT CLAIMED? 126

IS THIS A GOOD MEASURE OF IMPACT? 126

HOW IS THE CLAIM SUPPORTED? 127

What Evidence Is Reported? 128

What Is the Quality/Strength of the Evidence? 129

IS THE CLAIM REASONABLE IN ITSELF?
DOES PRIOR BELIEF IMPACT MY BELIEF?
CONFIRMATION BIAS 129

COMMUNITY WITH LOW RATE OF INFECTION 129

Rapid, Less Accurate Test 129

Slower, More Accurate Test 130

COMMUNITY WITH A HIGHER RATE OF
INFECTION 130

Rapid, Less Accurate Test 131

Slower, More Accurate Test 131

HOW MUCH DOES THIS MATTER TO ME? 132

CONSIDERING THE COVERAGE 132

REVIEW 134

STATS + STORIES PODCASTS 135

CHAPTER 9 ▪ Research Reproducibility
and Reporting Results 139

STORY SUMMARY 140

WHAT IDEAS WILL YOU ENCOUNTER IN THIS
CHAPTER? 142

WHAT IS CLAIMED? IS IT APPROPRIATE? 142

WHO IS CLAIMING THIS? 143

WHY IS IT CLAIMED? WHAT MAKES THIS
A STORY WORTH TELLING? 143

IS THIS A GOOD MEASURE OF IMPACT? 144

HOW IS THE CLAIM SUPPORTED? 144

What Evidence Is Reported? 144

What Is the Quality/Strength of the Evidence? 147

DOES THE CLAIM SEEM REASONABLE? 149

HOW DOES THIS CLAIM FIT WITH WHAT IS
ALREADY KNOWN? 149

HOW MUCH DOES THIS MATTER? 150

Comparison of Population Perspective versus
Individual Perspective? 150

Will I Change My Behavior as a Consequence of This? 150

CONSIDERING THE COVERAGE 151

REVIEW 154

CODA: A NEW 3 R'S? 156

STATS + STORIES PODCASTS 157

CHAPTER 10 ■ Now, What? 159

CONSIDER THE WEIGHT OF EVIDENCE 163

CONSIDER THE SOURCE 166

CONSIDER THE HISTORY 166

BE A CRITICAL READER … OF EVERYTHING 167

BIBLIOGRAPHY, 169

INDEX, 173

Preface

(OR THE PART OF a book that many readers ignore even though the authors spent time thinking about it – this may tell you if you'll like the book and the authors!)

TLDR: WHY THIS BOOK AND WHY NOW?

Good decisions need good evidence. We need skills to understand data, how it is measured and how it is analyzed and how it is presented to yield information to support action. Statistical literacy can be a superpower for extracting information from data. Learning statistics is best motivated by stories, and the story before concepts and structure approach is used here. Understanding what reporters and editors are thinking may help you become a more critical consumer of the news. It may also help you think about how you tell your own stories about statistics – what do you need to ensure someone understands and why?

Colleagues in official statistics say that a good government needs good data. Our belief is that an informed citizenry needs a good understanding of how data are generated and used as part of narratives to be able to critically consume this information. The role of journalists is to ensure that individuals have access to such information; understanding how they frame information can help you understand why particular stories circulate while others do not. This book is intended to empower people to deconstruct the headlines and stories and to investigate and explore whether claims are justified or not.

WHY NOW? (POST-COVID-19 PANDEMIC ADDENDUM)

We started seriously working on this book in the Spring of 2020. Months into this new year, we learned that the words of the year would be "unprecedented" and "surreal" as the result of the COVID-19 pandemic. Projections of the course of the number of cases and fatalities were being made early in the spring, and output from these model-based projections was being used to support control measures such as social distancing. Comparing distributions of the number of cases under different scenarios of exposure led to the entire world looking to "flatten the curve" to avoid overwhelming our healthcare systems. Exponential growth and doubling times were now routinely reported. A confidence interval was even reported at a White House press briefing. Uncertainty in models, describing and visualizing data, variability in properties such as infectivity before displaying symptoms, and sensitivity of new tests for the virus were not esoteric topics to be sequestered in a statistics class but now ideas that many citizens strongly desired to understand.

Journalists had a hand in shaping that understanding. Newspapers were filled with graphics showing how curves might be flattened, describing the way the novel coronavirus might spread in an enclosed space, and detailing the ever-growing list of symptoms an individual with COVID-19 might exhibit. As the public found itself awash in data, so, too, did journalists. Famed for a mythic avoidance of numbers, journalists now found themselves working to make sense of positivity and reproduction rates so that they could meaningfully report on the pandemic for their communities. The data that drives so much decision-making – but which is generally crunched behind the scenes by policymakers or other experts – was front and center. Now, more than ever, there is a need to develop statistical literacy to think critically about the headlines and their associated stories.

WHO IS THE TARGET AUDIENCE FOR THIS BOOK?

This book is intended for a general audience. If you can read and you are interested in the world in which you live, then you've met the prerequisite background. If you are a consumer of news that is based on some background foundational work, then this book is for you. We also believe this book would be a complementary and supplementary resource for quantitative literacy courses, introductory statistics and journalism courses and seminar courses of all types.

WHY DO WE THINK STATISTICS CAN BE LEARNED FROM THE NEWS?

While for the rest of this book we will be speaking collectively, we'd like you to allow us this minor digression into our individual perspectives on why we feel this matters.

John's Perspective

I have a confession. I teach statistics. Self-identifying as a statistics teacher would be a great way to stop a conversation at a party many years ago (maybe this was just me but my friends confess to similar receptions). Early in my teaching career, I would bring stories from the news (I would cut out stories, yes with scissors, from newspapers and would then display them on an overhead projector!) to engage students in conversations about what was being claimed in a headline and story, what data was* collected and how it was analyzed and presented. Students responded and engaged with these stories; this simple exercise appeared to open their hearts and minds to think about the statistical thinking needed to understand these stories. (* data was/data were – I used to get agitated about "data" being plural versus "datum" singular so "data were" was non-negotiable. I've mellowed with changing common use.) Data *is* when we talk about data in the sense of "dataset." Data *are* when we talk about a collection of individual observations. Also, language is fluid and it evolves. We don't use words the same way they were used 100 years ago.

Cohn and Cope (now with Cohn Runkle) wrote a brilliant book, *News and Numbers*, that connected the questions that news stories with significant statistical content needed to answer. While that book was intended to support journalists interacting with research and study design, it also provided a series of questions that a researcher needed to be ready to answer for a journalist. I used these questions as a motivation for what issues needed to be addressed in introductory statistics classes. More importantly, this changed how I thought about communicating research and study results to the public.

Team-teaching, a News and Numbers course with a journalism colleague, Richard Campbell, reinforced my earlier practice but also challenged the way that I think about statistics and how I consider motivating people to learn statistical thinking. We used the Cohn and Cope book along with Joel Best's *Stat Spotting: A Field Guide to Dubious Data* as texts as background texts, and we required students to produce portfolios of statistical practice and stories with statistical content. The story behind the statistics was the usual motivation for reading a story, and the statistics behind the story often provided the foundation for the story.

Rosemary's Perspective

Here's my, perhaps somewhat sacrilegious, confession: I once was an avowed number hater. I found math and statistics incredibly boring. I would rather help my father, who was a contractor, deal with a plumbing issue than work on my algebra homework in high school or my statistics homework in college. And then I graduated from college, began a career as a journalist, and realized I was passionate about telling science and medical stories. Do you have any idea how difficult it is to brush up on statistics while working in a newsroom fulltime? Frankly, I'd rather have my wisdom teeth pulled out than do that again. But, here I am, writing this book after a journalism career that saw me win a couple of science and medical writing awards. Not too bad for a number hater.

The thing is, I don't hate math or stats now. In fact, in graduate school, I took a statistics class where the professor made me run chi-square and linear regressions by hand – and I found that I actually loved it! For me, it took having a teacher who could tell me the story of what these stats do that finally got me hooked. Now in my research, I happily straddle the quantitative–qualitative line, allowing the questions I'm interested in answering to help me decide on my method. Had someone not helped me see the beauty in statistics and what they can help us understand, I would not be able to do the kind of work that I care about as a scholar.

My job as a journalism professor is to try to help my students let go of their own fear of numbers – it is truly a fear many a journalism student has – and help them see how becoming statistically literate can help them do their jobs better and better serve the public. Journalism is not done in a vacuum; the purpose of the institution is to ensure the public are informed enough about their world that they can make informed decisions about their civic lives or public health. While my own push to revisit statistics as a reporter came from wanting to be able to tell stories about HIV/AIDS or cancer treatments or beer brewing well, statistics are also important in stories about economics, city hall and the judicial system. Being statistically literate opens up data and information to reporters that allow them to tell complex, contextualized and nuanced stories. That information appearing in news stories can then help readers, listeners and viewers make informed decisions about their lives. Statistics can be communicated well in news stories; it just takes time, care and effort. As does the podcast John and I appear on.

FROM TEACHING TO TALKING TO WRITING

After teaching the News and Numbers class, we (John and Richard Campbell) cooked up the idea for another collaboration – a podcast with a name that reflected the same parallel connection that we see in News and Numbers – Stats + Stories. Since our inaugural

release in 2013, the Stats + Stories podcast (www.statsandstories
.net) has worked to explain the statistics behind the stories and the
stories behind the statistics. In 2015, Rosemary joined the podcast
as the moderator and, after years of doing the podcast, the time
felt right to bring statistics and stories together as the foundation
of a book to introduce statistical and journalistic concepts. Hence,
the text you now hold in your hands (or are reading on your screen
or giving as a heart-felt gift to someone you love).

STRUCTURE OF THIS BOOK

David Spiegelhalter, statistician, former Royal Statistical Society
president and three-time guest on the Stats + Stories podcast,
presented about the structure of how science moves from the
technical literature to a headline in a forward path at both the
International Statistical Institute World Statistics Congress in
2017 and in Figure 12.1 of his excellent 2019 book, *The Art of
Statistics*:

Scientific publication/government report => press release
=> story => headline.

A question one step removed is what are the critical statistical
ideas important for understanding and critically consuming the
information in the scientific publication/government report? A
complementary question is what are the features of the research
or report that make it newsworthy?

The approach in this book is to invert this path to use
the headline and story to motivate an investigation into the
background scientific work and to introduce statistical and
journalistic ideas as needed to evaluate the story. In other words,

Headline => story => press release => scientific publication/
government report => what statistical and journalistic concepts are needed to understand the story?

The structure of each chapter: headline/story/background material/concepts is a natural framework for a critical reader to apply to consuming news or for teachers to apply in a homework assignment.

In this book, a chapter will start with a headline (often click-bait for readers – it was probably for us if we included it in this book) and will then

- Consider whether the details in the article supported the headline and whether the scientific paper or government research report supported the article and its headline.

- Using a series of common framing questions, review the scientific paper behind the story to:

  - Determine if the story and headline were reasonable, overblown or incorrect.

  - Introduce statistical concepts needed to critically consume, or at least appreciate, the research.

  - Examine the journalistic concepts associated with the reporting in this story.

The stories/chapters were selected to span a range of topics including economics, education, entertainment, environment, health, media, politics, society and sports. In fact, many of these stories will hit a couple of topic categories. The first chapter provides more information about the framework and structure of the following chapters and provides a table of topics covered in each chapter. While we encourage you to skim this chapter first, feel free to dive into the chapters that follow in any order that appeals to you.

Finally, a GitHub repository, https://github.com/baileraj/StatsBehindTheHeadlines, includes code for figures and any analyses in the book along with data.

Acknowledgments

JOHN'S ACKNOWLEDGMENTS

I thank John Kimmel of CRC/Chapman & Hall for his encouragement to consider contributing to the ASA Practical Significance book series. I also blame him for the work that followed in the wake of the book proposal being accepted. I had the pleasure to work with John K. on a number of books, and this experience was no different. After John's retirement in 2021, Lara Spieker became our editorial muse, and we are grateful to her for bringing this book train into the published station.

I also thank Sir David Spiegelhalter for many conversations and for his generosity in hosting a short visit to the Winton Centre for Risk and Evidence Communication at Cambridge University in early 2020. David, the Winton Executive Director Alex Freeman and colleagues there (María del Carmen Climént, Sarah Dryhurst, Claudia Schneider, Holly Sutherland, Alice Lawrence, Gabe Recchia, Ilan Goodman, Mike Lawrence and Jin Park) were very welcoming, and the conversations with this team helped frame the questions addressed in each chapter. David and Anthony Masters have produced an amazing column in *The Guardian* and an accompanying book that provide amazing examples of communicating statistical concepts related to the COVID pandemic, vaccines and more.

I am grateful to Miami University for the research leave that provided the time for me to work on this project and for the support of collaborations. I work with great colleagues in an energetic department with the support of colleagues throughout the university.

I already thanked the larger community of trailblazers in data journalism and statistical issues in journalism; however, my personal and local thanks go to Richard Campbell, my oldest partner-in-crime in this work of communicating the statistics behind the stories and the stories behind the statistics, and Bob Long, my now paroled partner-in-crime. I have learned a lot from these journalism colleagues about asking good questions and listening with focus. My final-journalism-colleague thanks go to Rosemary Pennington, my coauthor on this book and continuing partner-in-crime on the podcast. I have learned much from her approach to broadcasting and podcasting and her framing of stories.

The Stats + Stories podcast is a labor of love that has been a product of this journalism–statistics partnership. I have learned much from the many guests we've had on the Stats + Stories podcast, and I am most grateful for their time and insights. In addition, I am grateful to the American Statistical Association for their support and partnership in promoting and expanding the work in this podcast.

Risk-taking is supported by having a strong foundation. My parents and grandparents encouraged inquisitiveness and modeled hard work, and my nuclear family powered my efforts to follow and develop my interests. (Yes, *nuclear* and *powered* were linked on purpose. I imagine my oldest daughter Sara would accuse me of making a dad joke at this point, and my colleague and coauthor is no doubt shaking her head in disgust.) Thanks Jenny for your support as I once again embarked on writing another book. I won't swear this will be my last.

ROSEMARY'S ACKNOWLEDGMENTS

I would also like to thank John Kimmel of CRC/Chapman & Hall for his encouragement as well as his flexibility with deadlines.

I'd like to thank Richard Campbell, my former chair in Miami University's Department of Media, Journalism & Film, for encouraging my involvement with Stats + Stories, and my current chair Bruce Drushel for his continued support of the podcast.

My colleagues in Miami's Journalism program have been a source of continued support, as well as interview questions, for the last several years, so a big thank you to them. Thank you, as well, to the Mongeese for the laughs, the groans and the memes and for just generally helping me keep my head on straight over the last year, as have Jessica von Ahsen Birthisel, Spring-Serenity Duvall, Lori Henson and Stacie Meihaus Jankowski.

When I was a kid, my dad would drag home piles of old National Geographics from the used book store, knowing that I'd read them cover-to-cover in a matter of hours. His encouragement of my voracious curiosity is why I became a journalist. I will forever be thankful for his love and care. I miss him every day. My mother patiently sat with me as I cried over the first D I ever got on a test – a math test – and told me that I'd do better next time. Her patient and unceasing cheerleading has gotten me through a number of challenges – including this COVID year – so, thank you, Mom, for never giving up, even when I wanted to.

I write this as my husband Tim and daughter Sofia are out running errands – partially out of a need to get out of the house, but also to provide me a quiet place in which to finish this writing. I'll never forget Sofia, when I was in grad school, asking me why I had so much homework all the time. Now the question is why I am writing all the time. I am so grateful for their support, love, kindness and the snacks they bring me when I forget to eat.

Finally, I have to thank my coauthor John. I should have never said yes to writing this, but I'm glad I did. It's been an interesting year.

JOHN AND ROSEMARY'S FINAL ACKNOWLEDGMENT

We would like to thank Alberto Cairo and Lynette Hudiburgh, readers of early draft chapters of this book, and the reviewers who provided feedback as part of the formal review process. This book is better for the incorporation of their comments and suggestions.

Our final thanks go to you, the reader. Thank you for your interest in exploring ideas that provide the foundation for good statistical thinking. We are delighted to have you accompany us on this exploration, and we hope that this book might provide encouragement and a foundation for you to explore the statistics behind the headlines.

REFERENCES

Best, J. (2013). *Stat spotting: A field guide to spotting dubious data* (1st ed.). University of California Press.

Cohn, V., Cope, L., & Cohn Runkle, D. (2011). *News and numbers: A writer's guide to statistics* (3rd ed.). Wiley-Blackwell.

Spiegelhalter, D. (2019). *The art of statistics: How to learn from data.* Basic Books.

Spiegelhalter, D., & Masters, A. (2021). *Covid by numbers: Making sense of the pandemic with data.* Penguin Random House UK.

CHAPTER 1

A Field Guide to Reading the *Statistics behind the Headlines*

Photo by Obi Onyeador (from unsplash: https://unsplash.com/photos/
UEQvUtRs224)

DOI: 10.1201/9781003023401-1

JOURNALISTS AND STATISTICIANS SHARE SIMILAR GOALS

The job of the journalist is to make the significant interesting. – Bill Kovach and Tom Rosenstiel's *The Elements of Journalism.*

Reporters should not lead their reports with a lot of numbers and data but rather tell a story that grabs the reader and illustrates the big data to come later in the story. (Richard Campbell's summary of comments from Nicholas Kristoff)

This is a book about statistics and the stories we tell about them. Journalists and statisticians often share the same goal: communicating complicated information to non-specialist audiences. Though, perhaps, statisticians are less concerned with making something interesting than accurate, like journalists they want their audiences to understand why particular numbers matter. Both journalists and statisticians work to make sure the answer to the "So, what?" question is clear. This book examines the methods used to help frame that answer with numbers and then considers how it is interpreted in the news. Ultimately, this is a book for all of us who care about how statistics are produced and used *and* the stories told about them. In the chapters that follow, we start with stories, consider whether the underlying work was significant and see what concepts are needed to put facts in context.

You have a choice as the reader now. Either be a traditionalist and read this chapter first because all books need to be read in order or rebel and don't read this chapter now. Both traditionalists and rebels are welcome, and this book encourages either reading strategy. The traditionalist will get a foundation for the structure of the book chapters and an overview of statistical concepts and some of the journalistic concepts that will be encountered in future chapters. The rebel will dive into a headline that looks interesting, read that chapter first and will be able to infer the structure of the chapter. At a minimum, both reader types might want to take a quick look at the text box describing the structure of later chapters and the table of methods by stories at the end of this chapter.

STRUCTURE OF EACH CHAPTER

Like all news stories, the following chapters start with a headline – one we found interesting and that made us want to read the article in the first place! More importantly, we thought there might be some statistical and journalistic nuggets worth mining in the story. We then work to excavate those nuggets with you, working through the statistical concepts which appear in the story before discussing journalistic practices that provide a foundation for a deeper understanding of the headline and the news story itself.

Unlike an unexpected finding after running a statistical analysis, we try to

1. What is claimed? Is it appropriate?
2. Who is claiming this?
3. Why is it claimed?
4. Is this a good measure of impact?
5. How is the claim supported?
   • What evidence is reported?
   • What is the quality/ strength of the evidence?
6. Does the claim seem reasonable?
7. How does this claim fit with what is already known?
8. How much does this matter to me?
   • Comparison of population perspective versus individual perspective?
   • Will I change my behavior as a consequence of this?

be upfront about the concepts we are focusing on in each chapter. We introduce them at the beginning and then do our best to point back to them throughout the chapter. While the same concept might be important in more than one chapter, any in-depth discussion of a concept will be restricted to a particular chapter. Like a good, inverted pyramid news article, we try to keep the writing concise and to the point (with full recognition we may fail in this endeavor at times).

Each chapter will review the story, *and* the journal article/ report on the story is about using a set of questions (see the text box). While the headlines initially caught our attention as we were reading the news, they were selected for inclusion in this book to allow us to discuss a broad array of issues while also reviewing particular statistical and journalistic concepts that are important for understanding these stories. Chapters will close with a recap of the story and, if you've not had enough of us, with links to Stats + Stories podcast episodes that have some connection to each story.

STATISTICAL CONCEPTS TO BE EXPLORED

Concepts are grouped into three broad areas: producing data, describing data, and drawing conclusions from data and then touching on fundamental concepts of causality and what is being measured. Kudos to Moore and McCabe for their book, *An Introduction to the Practice of Statistics*, for influencing and framing our thinking about the three broad areas.

WHAT IS IT? HOW MUCH IS THERE?

Not everything is easy to define or measure. While we are relatively confident about physical measurements such as distance and weight measurements; however, defining and then measuring happiness or pain or intelligence may need more work in developing the variables and validating scales for these measurements. These concepts are explored in the chapters that follow.

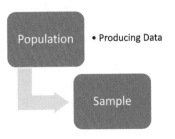

DATA GENERATION/PRODUCING DATA

Data doesn't grow on trees. (Well, some forestry scientists might argue with us about that.) Data arise from some process of cycling through a motivating question followed by observation and measurement. The motivating question determines who is observed and what is measured. Observations may be produced by sampling, experiments and working with existing data sources such as using application programming interfaces (APIs) or scraping from web pages. Data might even be simulated in a computer with no physical reality. Later chapters will describe how each of these processes works.

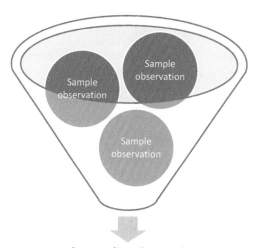

Summary (Describing Data)

DESCRIBING DATA

Before data can be transformed into information and insight, we need to explore it and analyze it. The description of raw data involves understanding the features of the pattern and the frequency of values of a single variable (aka distributions) and the relationship between variables. The numerical and graphical summaries of distributions and relationships need to reflect how variables are defined and measured. You'll also discover what "risk" is and how "risk" might be defined. In the upcoming chapters, we'll explore these topics and many more.

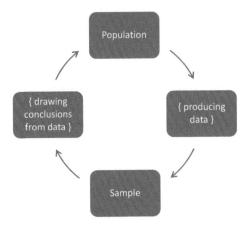

DRAWING CONCLUSIONS FROM DATA

What is the relevance of data and conclusions from a particular study or report for some broader population? Which population? Can we estimate features of any population based on insights obtained by examining a small sample drawn from that same population? Can we formally evaluate hypotheses about this larger population based on the study results? In addition to these questions that are part of inferential statistics, we want to consider how variables might be connected for prediction and classification in models. Consideration of uncertainty and variability overlays this discussion.

Outcomes in the real world may be determined by a host of input factors that interact. The collection of interconnected inputs and outputs is not fully known. Uncertainty reflects the unknowns about this collection such as inadequacy of our understanding of this collection. In addition, variability is intrinsic in most things we study. Usually, uncertainty about this collection can be reduced with more data and more theoretical understanding, but variability will be better described. Uncertainty has been described as a characteristic of a researcher and variability as a characteristic of a system. We think of both as being job security for statisticians, data analysts and data journalists, and source of rich material for reporters.

IF I DO THIS, THEN THAT WILL HAPPEN

Finally, causality is often implicitly suggested in headlines and stories even when not suggested by the background research or by the investigators doing that work. Even if not intended by the story, readers may jump to assumed causal relationships. Our imposition of causal relationships may be part of our desire to explain the world. In each chapter where it makes sense, we consider whether causality was suggested in the headline, story or original source. In addition, we consider if other factors may have impacted the observed relationship.

JOURNALISM 101

Our framework for discussing the statistical concepts at the heart of this book is news stories about statistical research. One of our goals in doing this is to help you not only understand how statistics are reported in news stories, but also to give you insight into why stories are reported and framed in particular ways. It's important to note that journalists are trained to be skeptics and taught to challenge authority, verify facts and investigate the veracity of claims. As you read about the news stories in this book, and continue reading news stories in the future, we hope that you will keep that idea of journalists as a skeptic in mind. It might help you understand why a story is reported the way it is; it might also help you understand where journalists fail in their coverage of something.

A news story starts in the same way that a study does – the reporter conducts background research into an issue to decide what the story is about, what the audience needs to understand and whom to talk to. The reporter then begins working to verify the facts of a story, consulting primary documents and interviewing sources to do so. As a reporter does this, they are also weighing the newsworthiness of a particular bit of information. The old adage in journalism is "if it's not new, it's not news" – so reporters are continually working to understand what is *new* in a story. Sometimes, it's a perspective that hasn't appeared in reporting on the topic in the past; other times, it's new data that help us understand something a bit better. That timeliness, however, is just one of many news values reporters consider when reporting a story. The impact of a story, the individuals it involves or its novelty can also influence a reporter to cover something.

Journalists are, generally, generalists. As you read about the news coverage of statistics in the book, think about how being a generalist might impact the way a news story is covered. What might the pros and cons of a generalist approach to covering this information be? What perspectives might be left out and how does that impact what you, as a reader, understand about what is being reported? For those of you who might be in the position to work with journalists in the future, we'd like you to think about how you'd approach communicating statistical information to reporters. If journalists are generalists, what do you absolutely need to make sure they understand to be able to report on your statistics well? Recognize that journalists as a general reader may serve as a surrogate for you – a nontechnical reader who is translating technical work because of the interesting story embedded therein. What role might researchers and statisticians have to play in the underreporting of some stories? How can you communicate uncertainty to journalists in a way that does not undermine your findings? And, journalists, how can you work to ensure that you understand the statistical information you are reporting on?

Communication is at the heart of what both journalists and statisticians do, though we may not share a common language for doing so. As you read the chapters in this book, we hope you'll think about not only the statistical concepts but also about how to effectively communicate them to your friends, your mom or that one person you know who says they just don't understand numbers. Whether you're a journalist or a statistician, learning how to clearly communicate complicated information is a skill that will serve you well after you've forgotten this book and its authors.

CHAPTER 2

Predicting Global Population Growth and Framing How You Report It

World's population could swell to 10.9 billion by 2100, U.N. report finds [subtitle] The growth will increase humanity's footprint on the planet, which could exacerbate hunger, poverty and climate change, experts say.) Story Source: NBC News (by Denise Chow) – https://www.nbcnews.com/mach /science/world-s-population-could-swell-10-9-billion-2100- u-ncna1017791

Government Report Citation: Population Division World Population Prospects 2019 – https://population.un.org/wpp/ Graphs/Probabilistic/POP/TOT/900

DOI: 10.1201/9781003023401-2

https://www.pexels.com/photo/grayscale-photography-of-people-walk-ing-in-train-station-735795/

STORY SUMMARY

Stories about populations are a journalistic staple, and this NBC News story about global population figures is a recent entry in this particular genre. In the article, reporter Denise Chow discusses the latest United Nations's (UN's) estimate of what the human population could grow to as well as how that growth might environmentally impact the earth. The report explains that the current population is estimated to be 7.7 billion; the UN's estimate for the population at the turn of the next century is 10.9 billion. The story then explains that, while this number is large, it is actually down a bit from what an earlier estimate suggested the population might grow to. The change, the reporter explains, is partly due to changes in the global fertility rate. People just aren't having as many babies as earlier estimates had projected.

What is novel about this particular story is its focus on what a growing global population might mean for a planet that is already reeling from the effects of climate change. The reporter also explains the "disconnect" between population growth and consumption – the places where the human population is growing are not the places where most of the consumption is taking place. The story ends with a reminder of the unequal way climate change will impact the earth's growing population.

WHAT IDEAS WILL YOU ENCOUNTER IN THIS CHAPTER?

- Models can be used to predict future events.

- Prediction is more uncertain for conditions far from where data are observed.

- Data source is a factor to consider when evaluating study results.

- Scope of impact can drive news coverage.

- Journalistic framing can determine how we understand a problem.

WHAT IS CLAIMED? IS IT APPROPRIATE?

As you evaluate the news stories you read that include statistics or statistical analysis as a major foundational focus, it's important to consider whether what the headline claims is actually backed up by what the story says. Most of the time, it will be. There may be moments, however, when it seems like you've experienced a bait and switch – the headline may have suggested one thing while the news story itself actually suggests something else. This kind of disconnect is seldom created with malicious intent. What is important to know is that the reporter who produced the story is rarely the person who writes the headline. Where the news story is designed to inform you about something, often providing context to do so, the headline's sole purpose is to make you want to read the story. You are the fish swimming through a sea of information, and the headline is the bait on the hook the news outlet is going to use to reel you in. In the case of this story about the earth's growing population, the headline and the story match.

While the headline features a projection of the world's population in 2100, the subtitle focuses on the impact of such an increase. The star of this first story is a statistic, the estimated size of the world's population. The claim in this story is that the population of the world will be 10.9 billion by 2100. Once the claim is made at the beginning, the reporter spends the rest of the story spelling out how researchers have come to this particular number, the context for it, and what a growing population means for the future of the planet.

Examples when a statistic is the star of a story:

- Population size
- Unemployment rates
- Taxes paid
- Vaccine efficiencies
- Virus infection rates

WHO IS CLAIMING THIS?

The source for this population projection is the United Nations Statistical Commission, which is part of the UN's Department of Economic and Social Affairs (ECOSOC). Many of the UN's other activities, such as General Assembly meetings, peacekeeping activities and the International Criminal Court proceedings, are well covered by news media. The work of ECOSOC, which produces statistics, receives much less coverage. At least, that's until it unveils things like the population projection that's covered in the NBC News story.

Reading Research:
- Do you trust the source?
- What is the reason the source produced the data and analysis?

Producing such projections is part of the remit and mission of ECOSOC. That's the kind of background you need in order to evaluate the soundness of the source of a particular bit of data. Questions you might consider include: Why did the data and study source produce this work? Do they have an agenda that is supported by the work that is being reported? Does the source have any financial or reputational gains that are associated with publishing the work? Does the source provide sufficient background that someone might use to reproduce or replicate the work? As an example, government-funded research typically makes data available for anyone to download and explore.

WHY IS IT CLAIMED? WHAT MAKES THIS A STORY WORTH TELLING?

The number is claimed because ECOSOC statisticians have spent a lot of time analyzing population statistics from around the world, working to contextualize them in relation to past estimates as well as in relation to what we know now about birth and death rates. The estimate, of a potential future population of 10.9 billion, is reported because of the possible impact of such a global population. Here, as noted by the reporter, the number of humans on the planet has implications for energy consumption, resource use, environmental impact and many other matters, and this is highlighted in the news story.

IS THIS A GOOD MEASURE OF IMPACT?

A well-reasoned guess of the size of the world's population is of intrinsic interest; however, the question of impact here is what this number might mean for future generations. The translation and interpretation of this number is the heart of the story as we can't know what immediate impact this predicted population growth might have.

HOW IS THE CLAIM SUPPORTED?

The claim is illustrated by a graphic where data from 1950 to 2020 are connected with a solid line and the future guesses with a dashed line that include a cone of uncertainty. So

Interpolation: A guess for conditions where data are available.
Extrapolation: A guess for conditions outside where data are available.

why make a distinction between the time periods 1950–2020 and 2021–2100? The numbers from the 1950–2020 interval are identified as "estimates" and in the later 2021–2100 interval are called "predicted population" in the news story and "projections" in the UN report. A prediction beyond the scope of your experience or available data is also called an **extrapolation** while a prediction within the scope of your experience is sometimes called an **interpolation.**

The predicted population trajectory is based on data from the UN website. A plot of the predicted population along with bands of prediction uncertainty is reproduced in the figure below.

The solid line is the median estimate of the predicted population and the endpoints of the shaded area correspond to an interval that reflects 80% of the uncertainty in the prediction process. The key conclusion from this display is that the size of the world population is predicted to grow over the next 80 years (the solid line); however, the uncertainty in the prediction also grows as reflected in the growing width of the 80% prediction interval (the gray cone around the center line). For example, this means that

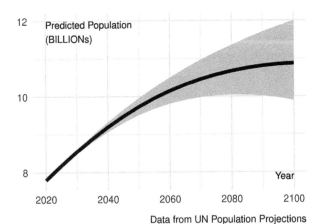

Data from UN Population Projections
(https://population.un.org/wpp/Download/Probabilistic/Population/)

FIGURE 2.1   Prediction of world population (solid line) with 80% prediction intervals.

four out of five model-based predictions of the world's population in 2100 are expected to be between 10 billion and 12 billion while four out of five model-based predictions of the world's population in 2030 are between 8.5 billion and 8.6 billion.

> Median: Estimate where half the values are expected to exceed this number and half are expected to be less than this number.
>
> Prediction interval: Range of values thought to be plausible for some unknown quantity.

This is consistent with your intuition where you are more confident about guesses in the near future, here nine years from now, versus guesses about the far future, say 80 years from now. Do you have more confidence in tomorrow's weather forecast than you do in the long-range 10-day forecast? It is important to note that this also means that one out of five model-based predictions of the world's population in 2100 is expected to be between either less than 10 billion or greater than 12 billion!

What Evidence Is Reported?

The original research reported predictions from a statistical or mathematical model that predicts annual population sizes

> Model: Representation of a system where **outputs** of interest are related to specific **inputs**.

for different countries or regions in the world which are then summed up into an estimate for the world.

While there is a complicated model operating in the background, its purpose is easy to understand: produce a prediction of the population size in any given year. What influences the size of a country's population in any year? Population changes from the previous year as a result of increases in births (aka fertility) and immigration and decreases in deaths (aka mortality) and migration out of the country. If you really want a formula for a country, then:

$$\text{Population next year} = \text{population this year} + \text{babies} + \text{immigrants}$$
$$- \text{deaths} - \text{emigrants}$$

The population of the world is determined by adding up all of the country-specific populations.

What Is the Quality/Strength of the Evidence?

While it is easy to describe the components that determine population size in a given year, it may be trickier in practice to determine those inputs. For example, birth rates and death rates differ for different ages, and further, these rates may be changing over time. So, any model for predicting population in the future needs to have estimates of how the birth and death rates are changing in the future. The number of immigrants into a country and migrants out of a country also may change over time. Even if this is fairly stable at the time the model is built, mortality may unexpectedly change with the emergence of new pandemics such as COVID-19 in 2020 or immigration/migration may change as a consequence of regional conflicts.

If you want to learn more about the model, details of the methods and the inputs used in the UN population projections can be found in a 60-page report (https://population.un.org/wpp/Download/Probabilistic/Population/). The report clearly defines the challenges of producing these estimates. Sections include fertility estimates – "Levels and trend of future fertility: convergence towards low fertility" and "Projection of age patterns of fertility" – mortality estimates that capture longer life expectancy patterns in all countries and age pattern assumptions and, finally, migration assumptions. All of these estimates have to be combined to produce annual population projections for 235 countries or areas and the world.

DOES THE CLAIM SEEM REASONABLE?

Expecting growth of the world's population seems reasonable and unsurprising. In addition, it makes sense that the population ultimately may level off. Growth in any system can only continue until the resources needed to promote growth are exhausted. This is sometimes called the carrying capacity of a system. To investigate models for growth that are unconstrained versus constrained, use your favorite search engine to learn more about "exponential growth" versus "logistic growth." You may have become more experienced with the concept of exponential growth as a result of discussions of how viruses can race through a population.

Our intuition is that the world's population is increasing. We didn't really have any strong feelings about how quickly the population was growing. The part of the analysis that grabbed our attention was the prediction that population would start to level out at about 11 billion residents. The rate of growth was projected to slow over time and appears to stabilize.

HOW DOES THIS CLAIM FIT WITH WHAT IS ALREADY KNOWN?

This is the 26th time the UN has generated these projections. Model inputs used to project the population in the future are calibrated/tuned using model-based estimates of the population size in years where other data on world population size is available (1950–2020) in this recent report. These inputs are then used to generate projections for the future. Recent trends in some of the inputs in the model such as trends in fertility are integrated into the analysis.

HOW MUCH DOES THIS MATTER?

How much and why a result matter reflects the result from scientific publications or in a government report that makes the story newsworthy. In this story, even the lower limit of the 80% prediction interval, 10 billion people, represents a 25% increase in the size of the world's population compared to the 2020 estimated population. At a time when discussions related to climate change, air pollution and sustainability are common, a 2 billion to 6 billion increase in population over the next 80 years merits attention.

While we may not be around to check the projections (and surely the UN will generate many more projections in the next 80 years), this predicted world population size matters because of the resources and environmental impact associated with it. The world's resources are finite and this UN project suggests population growth is likely to level off. But will it level off quickly enough to mitigate major environmental impacts?

Comparison of Population Perspective versus Individual Perspective?

While this chapter focuses on the projection of the world's population, the country-specific and regional projections might be more useful for the implications for resource utilization. For example, if you live in a country that was already experiencing increasing demands for freshwater for home and agricultural uses and your country was projected to have population growth in excess of 25% over the next 80 years, then you might feel even stronger pressure to act to reduce and recycle water use. One of the things a reporter tries to do when telling a story about statistics is to help the audience understand how the big gestalt figure may impact their lives.

Will I Change My Behavior as a Consequence of This?

If you are motivated by a sense of legacy and the call to leave a place better than you found it, this population projection is a little scary. This projection, in combination with the knowledge you might possess prior to reading this story and reviewing the UN report, suggests that demands and stressors are likely to increase. Actions to promote good environmental stewardship will likely increase as a consequence of this projection as will perceptions that demands on the world's resources will be markedly impacted even at the low estimate of the 2100 population. If you haven't thought about limited resources or how large populations impact the environment, then your reaction to the story might well differ from what is described here.

Having said this, it is worth noting that communication scholarship has generally shown that people don't change their behaviors based on a single story they read, especially if it happens to produce feelings of guilt or shame. In addition, articles consistent with prior belief might be read with enthusiasm while those that counter prior belief might be met with skepticism or outright denial. So, while John's behavior might change, many of our behaviors won't. Now, if news outlets continue covering the issue, adding more context as well as reinforcing the points made in this story, that might convince some people to change their habits, particularly if those stories create an emotional response in the audience. The emotion most likely to get us to do something? Anger. (As an aside, this is also the emotion that may lead to higher response rates in voluntary response samples.)

CONSIDERING THE COVERAGE

Stories about populations are a regular feature of journalism. Scholar Alison Hedley (2018) points out that what she calls "population journalism" was particularly popular during the Victorian era, driven in part by the use of colorful data visualizations. The problem with covering stories about human populations can be finding what journalists call the "news hook" – the bit of new information that will draw an audience into a news story. Here, the hook is not only the bare figure – 10.9 billion – but also the implications of that population boom on the planet as a whole.

One of the news values discussed in Chapter 1 was "impact" – basically, it's the idea that one way to get an audience to care about an issue is to explain the effect it will have on them. While

> Signals (or Symptoms?) of Bad News Reporting:
> Be on the lookout for either
> 1. Oversimplification or
> 2. Overcomplication.

the 10.9 billion people are estimated to be here 80 years into the future, NBC's tying this figure to the issues of resource depletion and global warming was a savvy way to quickly communicate impact. Why should audiences care about this somewhat abstract number? Because we or our loved ones are going to feel the crunch as the population grows.

Something this news story does well is explain how the population will continue to grow even as fertility rates around the globe continue to decrease. One of the hallmarks of bad news coverage of statistics is writing in such a way that either (a) oversimplifies the data or (b) overcomplicates it. (Which can be a problem for researchers writing articles about their own statistical analyses as well.) Either would be easy to do when juggling the contradiction of decreasing fertility and increasing population. After talking through the numbers, though, the reporter explains "The revised figures reflect the downward trend in the global fertility rate, which means the populations of more countries are shrinking," before telling the audience what specific areas of the world are expected to continue growing. It's a little thing that adds a lot to an audience's understanding of the issue. One challenge is that details and nuance can make stories more difficult to tell for the writer and to parse for the reader.

Perhaps what might have been handled a bit better is the discussion in the latter half of the story of how consumption habits impact a changing climate. The story quotes an expert on the differential consumption habits of people across the globe before shifting its focus to resource depletion. The reporter writes,

> In other words, the average person's lifestyle in the U.S. is more detrimental to the environment than the average person's in sub-Saharan Africa. That means rapid population growth in Africa won't be as damaging to the environment as a similar population increase would be in the U.S.

It might have been useful here to bring in some hard statistics that showed this disparity in a more concrete way rather than leaving it in the abstract. Putting that disparity before an audience in black and white figures could have helped communicate more concretely the urgency of this issue of overpopulation, while also helping the audience understand more fully that this is not a bare numbers game. The use of images such as those on Gapminder Dollar Street (https://www.gapminder.org/dollar-street) might be another mechanism to illustrate differences between countries' use of energy. For example, the images of sources of heat as a function of household income in different countries might be connected to fossil fuel use increases as populations increase in these countries.

Overall, this is a well-reported story, one that does not sensationalize the report even as it works to convince the audience of why they should care about an event so far into the future. Other outlets were not so careful in the coverage. One "pro-consumer" science advocacy organization called the UN report proof that overpopulation is a "myth" because the estimate was lower than that published in an earlier UN report.[1] An opinion writer in Australia suggested that the globe was headed for a "population crisis, but not the one it was expecting"[2] – and then launched into a discussion of declining birth rates. An Indian newspaper took a similar angle, suggesting that declining fertility rates will lead to a "demographic apocalypse"[3] as wealthier companies will compete for skilled migrants, which could change the demographic makeup of nations. While the three pieces referenced above fall more into the opinion category of journalistic publications, they are an important reminder that writers will often take statistical information and use it to frame an argument they want to make. In addition, this shows how a particular study may be the foundation of a number of different stories, including those with a flavor of opinion. What makes the NBC News story so solid is its discussion of a number of the main points in the UN report and its contextualization of them. When reading news stories about statistical information, that is what you want to see. Not every news story is going to be as long as the NBC story or will be able to cover as much, but it should be clear that the reporter is letting the data tell the story and not trying to shove the data into a story they wanted to tell.

REVIEW

The NBC News story discussed in this chapter reported extrapolations of future world population sizes based on a model that was developed and fit by UN population researchers. The news story captured both the predicted increase in world population and the increasing uncertainty associated with predictions farther into the future.

STATS + STORIES PODCASTS

Stats + Stories has featured a number of conversations about official statistics. Stefan Schweinfest, Director of the United Nations Statistics Division, discussed data used by the UN and efforts to make sure data collection efforts are supported and sustained – https://statsandstories.net/society1/the-un-and-statistics.

John Pullinger talked about the role and importance of official statistics, describing it as a human right (https://statsandstories .net/methods/statistics-are-a-human-right), and Gemma Van Halderen described official statistics in Asia and the impact of COVID-19 on official statistical practice (https://statsandstories .net/methods/official-statistics-in-asia-and-the-pacific).

The conflict that might emerge for an official statistician reporting estimates in conflict with the government is part of the conversation with Andreas Georgiou (https://statsandstories.net /economics1/2018/8/2/holding-up-a-mirror-to-society-a-tale-of -official-statistics-in-greece-stats-stories-episode-60).

The U.S. Census and controversies associated with it have been the focus of a number of episodes with guests Rob Santos (#160, #123 in https://statsandstories.net/episodes list), Hansi Lo Wang (#128), Chaitra Nagaraja (#116), Mike Annany and Mark Hansen (#106), John Thompson (#31, #32) and Tommy Wright (#2).

Other episodes with an official statistics twist featured guests Linda Young (#45 agricultural statistics), Barry Nussbaum (#20 environmental statistics), Jeri Mulrow (#35 crime statistics), Brian Mayer (#38 economic statistics) and Lisa Lavange and Ron Wasserstein (#75 controversy with moving a statistical agency).

REFERENCES – WORLD POPULATION PROJECTION

Report *World Population Prospects 2019: Methodology of the United Nations Population Estimates and Projections.*

https://population.un.org/wpp/Publications/Files/WPP2019_Methodology.pdf

https://population.un.org/wpp/Methodology/

NOTES

1. Overpopulation myth: New study predicts population decline this century, American Council on Science and Health. Accessed at https://www.acsh.org/news/2020/08/05/overpopulation-myth-new-study-predicts-population-decline-century-14953.
2. Peter Hartcher, The world heading for a population crisis but not the one it was expecting, The Sydney Morning Herald, July 21, 2020. Accessed at https://www.smh.com.au/national/the-world-is-heading-for-a-population-crisis-but-not-the-one-it-was-expecting-20200720-p55dl1.html.
3. Hugh Tomlinson, Demographic apocalypse: Collapsing birth rates will turn our world upside down. The Times of India, July 22, 2020. Accessed at https://www.thetimes.co.uk/edition/world/demographic-apocalypse-collapsing-birth-rates-will-turn-our-world-upside-down-bxpbjrbk0.

CHAPTER 3

Social Media and Mental Health

The Big Number: 3 or more hours a day of social media use hurts youths' mental health. Story Source: *Washington Post* (by Linda Searling) – https://www.washingtonpost.com/health/the-big-number-3-or-more-hours-a-day-of-social-media-use-hurts-youths-mental-health/2019/09/27/043c7dc0-e06f-11e9-8dc8-498eabc129a0_story.html

Scientific Journal Citation: Riehm KE, Feder KA, Tormohlen KN, Crum RM, Young AS, Green KM, Pacek LR, La Flair LN8, Mojtabai R. (2019) Associations Between Time Spent Using Social Media and Internalizing and Externalizing Problems Among US Youth. *JAMA Psychiatry.* 2019 Sep 11:1-9. doi: 10.1001/jamapsychiatry.2019.2325.

https://jamanetwork.com/journals/jamapsychiatry/fullarticle/2749480

DOI: 10.1201/9781003023401-3

Photo by mikoto.raw Photographer from Pexels. https://www.pexels.com/photo/photo-of-woman-using-mobile-phone-3367850/

STORY SUMMARY

We are somewhat obsessed with social media – whether it's the newest meme, the latest TikTok dance craze or how much time we are spending on Facebook or Instagram, we think about it a lot. But, that might not be all that healthy. Researchers have been working to understand the effects of social media use on our well-being, with many taking a specific focus on how social media use impacts the lives of children and teenagers. This *Washington Post* story is about a study that suggests more than three hours a day spent on social media is unhealthy for young people. The study sought to understand the impact of social media use on the mental health of young teenagers, with reporter Linda Searling pointing out that the research suggests too much use might be correlated with negative mental health outcomes. However, Searling tells readers that researchers aren't quite sure what the mechanism is that might lead to those outcomes.

WHAT IDEAS WILL YOU ENCOUNTER IN THIS CHAPTER?

- Defining and measuring variables are an important part of research.

- When and how data are collected are important for drawing conclusions about populations.

- Describing variables includes central summaries, variation and shape.

- Risk comparisons can be defined in different ways.

- Uncertainty should always be reported along with the best estimates.

- It is important to adjust analyses for all the variables that might impact the response.

- Causality in relationships is critical to evaluate.

- Scientific papers can be read strategically and key features extracted.

- News coverage can range from simple summaries of results to full expositions and explorations.

WHAT IS CLAIMED? IS IT APPROPRIATE?

The headline claims that social media use *hurts* the mental health of youth. This suggests that an activity (social media use) causes an adverse outcome (decline in mental health). This title highlights a common challenge of needing to unpack the implications of story headlines and later story components. One question you might ask is whether the term "hurts" – as used in the headline – is appropriate given the information that follows in the story. And, if it does not, then it's worth considering what may have driven a news outlet to characterize information in a particular way. Is it a

problem of oversimplifica-
tion? Or did the headline
writer not understand
the story or the study?
Ideally, as you read the
story, it should become
clear whether the headline
was appropriate or not. It
is always worth recalling
that the story writer and
the headline writer are
rarely, if ever, the same
person. Finally, consider
which of the two head-
lines would have enticed you to read the story:

Causality: Does some response
(here, mental response) result
as a consequence of some
activity (here, social media
use)?

Reverse causality: What if the
activity (social media use) is a
consequence of the response
(mental health condition)?
Here, what if mental health
status causes social media use
instead of the other direction?

- Three or more hours a day of social media use hurts youths'
  mental health.

- Three or more hours a day of social media use might be associated with youths' mental health, but we're not exactly sure
  why.

WHO IS CLAIMING THIS?

This was a health story in the *Washington Post* based on a scientific paper published in one of the journals of the American
Medical Association. An important distinction between the story
can be observed in the headline of the story and the title of the
scientific paper. The story headline implies that excessive social
media use hurts mental health while the scientific paper headline uses the word "associations." Causality is suggested in the
news story while more caution is embedded in the scientific paper
title. This difference is not an indictment of the inability of scientists to make definitive statements or the boldness of journalists
and editors to do so. This reflects a different set of priorities in

summarizing work: the scientists have the time and page space to describe a story with nuance and caveats to an audience that would expect no less while journalists and editors have much less space to attract and engage casual readers.

WHY IS IT CLAIMED?

Data were collected on mental health, social media use and other variables that might be associated with mental well-being. A statistical model was fit, suggesting that the odds of exhibiting some of the symptoms of adverse mental health increased with increasing social media use. Particular levels of social media use by 12- to 15-year-olds in the United States from 2013 to 2014 were associated with twice the likelihood of depression, anxiety, loneliness, aggression or antisocial isolation compared to their non-social-media using peers.

IS THIS A GOOD MEASURE OF IMPACT?

Impact is measured in terms of the odds ratio – a summary that reflects relative risk as described in the journal article – of mental health problems when comparing different levels of social media use. Two questions need to be explored here: (1) how do you measure variables such as mental health and social media use and (2) what do odds represent?

Variables

You might feel depressed or anxious or angry but how depressed, anxious or angry are you? Are you more depressed than you were yesterday? Are you more depressed than someone else? Can you determine these features with a set of questions? How will you know if the questions measure what you claim?

All studies require decisions about what to measure. In addition, there are often serious questions about how to measure these things. If we are measuring physical quantities such as distance or weight, then scales and units of measurement are clear. If we are measuring psychological characteristics or mental health, then scales need to be developed.

A few properties of scale development and measurement are **validity** and **reliability**. If you went in for a health screening assessment and your blood pressure was first 140/110 and then 132/95 and then 120/90 on consecutive readings, you might con-

Validity: Device (scale, questionnaire) used to measure the feature of interest (weight, depression) actually measures the feature.

Reliability: Device reproduces a measurement over multiple instances.

clude that their blood pressure cuff was not reliable. (Yes, one of the authors did make them take the measurement three times; yes, this author did suggest that they might want to recalibrate their blood pressure machine, and yes, they were glad to get rid of this annoying individual.) This digression opens up the question of whether the response is highly variable – blood pressure can be – or whether the measuring instrument was imprecise.

These traits or variables can be described using different scales of measurement (Table 3.1).

Scales of measurement are important because summaries or analyses of data need to reflect the scale of measurement. For example, if you record the height of players and their football jersey number, then you would have two numbers per player. It would be nonsense to calculate the average jersey number; however, the average player height might be an interesting and relevant summary.

Aspects of mental health in the study the *Washington Post* wrote about were measured using internalizing problems (e.g., anxiety and depression) and externalizing problems (e.g., aggressive behavior, bullying). Each participant completed a questionnaire which featured symptoms of these problems. Youth with four or more internalizing (externalizing) symptoms were classified as high with respect to internalizing (externalizing) problems. This questionnaire had five questions related to internalizing problem symptoms and five questions related to externalizing problem

TABLE 3.1   Common Scales of Measurement

Scale	Feature of Scale	Example
Nominal	Levels of scale differ in name/quality	Species detected in a survey of wildlife
Ordinal	Nominal + levels are ordered but the differences between levels are not interpretable	Pain scales: Describe your level of pain on a scale from 1 (no pain) to 10 (debilitating pain); teaching effectiveness scales from 0 (stinks) to 4 (amazing). Arguably, the difference between teaching effectiveness scores of 0 and 1 does not reflect the same difference in effectiveness as between 3 and 4 even though these are 1-unit differences on the scale
Interval	Ordinal + difference between levels is interpretable	Temperature (°C): The difference between 10°C and 20°C is the same as the difference between 37°C and 47°C
Ratio	Interval + true zero on scale so ratios between scale levels are interpretable. Here, zero on a scale means the absence of a trait	Length: A 5K race is 50 times as long as a 100 m race

symptoms. Three or fewer symptoms were labeled "low to moderate." An adolescent in the study was classified into one of four categories:

1. No problems.

2. Only internalizing problems.

3. Only externalizing problems.

4. Both internalizing and externalizing problems.

Finally, these measures are **self-reported**. Would a 12- to 15-year-old report bullying or other external behaviors that might be viewed negatively by others? Would you change your answers if you were asked a question that had a negative social consequence?

Most researchers will have a protocol to ensure anonymity to encourage honest responses.

Odds and Odds Ratios

You often read about odds associated with sporting events. An **odds** is the probability that something happens divided by the probability that something doesn't happen. **Probability** is a measure of the likelihood that

> Probability = likelihood something will happen
>
> Odds = Likelihood something happens/likelihood it doesn't happen
>
> Odds ratio = Odds in one condition/odds in different condition

something happens that ranges from 0/0% (certain to *not* occur) to 1/100% (certain to occur). Another word for probability commonly used in a health-related situation is **risk**.

If your weather app suggested that your hometown has a 60% chance of rain tomorrow, the odds of rain tomorrow can be calculated as:

Odds of rain tomorrow

$$= \{\text{Probability that it rains tomorrow}\} / \{\text{Probability that it does not rain tomorrow}\}$$

$$= \{\text{Probability that it rains tomorrow}\} / \{\text{1-Probability that rains tomorrow}\}$$

$$= 0.6/(1-0.6)$$

$$= 1.5 \text{ or } 60:40 \left[\text{or } 3:2 \text{ or } 1.5:1\right].$$

A few years ago, the pre-season odds of Leicester City winning the English Premier League (EPL) were 5000:1 against. This implies that the odds-makers thought that Leicester City has a 1 − (5000/5001) = 1/5001 = 0.0002 probability of winning the EPL. Imagine the surprise when an outcome of 2 in 10,000 was observed!

If the odds > 1, then this suggests that the event is more likely than not, and if the odds are < 1, then this suggests that the event is less likely than not. Suppose you are thinking about traveling to a city with a 30% chance of rain tomorrow. The odds of rain in this city are 0.3/(1−0.3) = 0.3/0.7 = 0.43:1 [or 3:7].

An odds ratio can be used to compare the odds of rain in your hometown to the odds at the city you might visit. Here, the odds ratio = 1.50/0.43 = 3.50, implying that the odds of rain at home are 3.5 times larger than the odds in the

> Relative risk = Measure of relative likelihood of an event under different conditions.
>
> Absolute risk = Direct measure of difference of the likelihood of an event under different conditions.

city you might visit. Note that this is a relative measure of impact. For example, if your weather app predicted a 10% chance of rain at home and a 3% chance of rain in the city you might visit (we know that most apps don't report such low probabilities, but just suspend your disbelief for a minute and play along), then this has a similar odds ratio, odds ratio = (0.1/0.9)/(0.03/0.97) = 3.59.

Odds ratios or other relative measures don't tell you the chance that something occurs, only how much more likely something is to occur in one situation versus another. In the first scenario (home rain chance 60%; away from city rain chance 30%), the chance of rain at home is 30% larger than the chance of rain in the away city. In the second scenario (home rain chance 10%; away from city rain chance 3%), the chance of rain at home is 7% larger. In both cases, the odds of rain are about 3.5 times high in my hometown versus the city I want to visit. In the first case, when the likelihood of rain at home is 60%, you are taking an umbrella. In the second case, you are not.

HOW IS THE CLAIM SUPPORTED?

Data were obtained after selecting a **sample** from a **population** of 12- to 15-year-olds in the United States from 2013 to 2014. Individuals were then followed for three years. A

> Population = Collection of all things (people, animals, countries) of interest.
>
> Sample = Subset selected from the population for study.

population is often too big to study in its entirety – how many 12- to 15-year-olds are in the United States? Do you even have the time and money to conduct a census of these adolescents?

Data can be produced in a variety of ways including **experiments**, **samples** and **observational studies**. This study uses a survey of adolescents in the U.S. population. It is nationally representative in the sense that each observation in the sample was selected with some known probability from the population. So, what?

If you know the probability that an individual was selected in the sample, then you can determine how many individuals in the population are represented by this sampled person, the **weight** attached to this person. Note that these probability samples can involve sampling at multiple stages. For example, you might first select states, then counties, then households and then youth from the households. The probability of the observation being selected is the product of the probabilities of selection in each of these stages.

One counter-intuitive observation is that the precision associated with estimates of population features depends on the size of the sample and not on the size of the population (assuming a sample has been

> Statistic: Numeric summary based on information in a sample.
> Parameter: Numeric summary of the population.

selected in some representative way). This observation is linked to how much you expect an estimate (**statistic**) of some population trait (**parameter**) to change from one sample to another sample from the same population. So precision is directly tied to sampling variability in probability samples.

Understanding sampling variability requires a thought experiment: imagine that you were going to do study after study where you sample 6500 adolescents using the same methods and estimate the average number of internalizing problems reported in each study. This average will vary from sample to sample and the amount that this will vary depends on the size of the sample, here 6500, not the size of the population being sampled. In contrast, this previous thought experiment with a smaller study where repeated samples of 65 adolescents were taken is considered. You might expect that any statistic calculated from samples of 65 would be more variable among repeated samples than the same statistic calculated from samples of 6500.

A potentially harder problem is that there may be other sources of error in a study. For example, what happens if your list of the population is not accurate? What if a sampled individual refuses to respond? Researchers conducting surveys work hard to reduce these non-sampling errors or attempt to adjust analyses when these are encountered.

In addition to starting with a sample, these adolescents were followed for an additional two years. Studies that collect data over time on the same individuals are called **longitudinal studies**. The measurements in the first year of this study (2013–2014) included age, sex, race, parent's education, BMI (body mass index), alcohol use, marijuana use and history of mental health problems. These so-called **covariates**

OBSERVATIONAL STUDY DESIGNS

Cohort: Follow individuals with particular characteristics and see what outcomes/responses are observed.

Case–control: Look at individuals with particular outcomes/responses and consider what characteristics they possess.

Cross-sectional: Grab a sample of individuals and simultaneously collect their outcomes/responses and characteristics.

represent variables that are thought to be important and possibly related to social media use. The second time period (2014–2015) was used to define social media use and the third time period (2015–2016) was used to define the mental health status. In this study, a "wave" is a window of time when measurements were

TABLE 3.2   The Waves and Variables Measured

Wave	Date Range	Variables Measured
1	12 September 2013 to 14 December 2014	Covariates include alcohol use, marijuana use AND history of internalizing/externalizing problems
2	23 October 2014 to 30 October 2015	Social media exposure
3	18 October 2015 to 23 October 2016	Mental health outcomes

taken, and the waves and the variables measured are presented in Table 3.2.

This data collection design has appeal in that social media use, the factor of primary interest (risk factor), preceded the mental health outcome measurement. While BMI, alcohol use and marijuana use appear only to be measured at the first time point, it would be no surprise if these characteristics change with age. Another restriction in the data used in this study is that individuals had to have data for all time intervals for inclusion in the analysis. What happens if an adolescent is more likely to drop out at later times because of poor mental health? Note that *all* studies have some limitations. This doesn't make the study invalid or useless; it simply acknowledges that there are a variety of factors to consider when evaluating a study.

What Evidence Is Reported?

Many research papers, including this paper, start results sections with summaries of the participants in the study and summaries of the responses of these participants. A

> Reading Research:
> Check out the first table of the article for a description and summary of the characteristics of the participants in the study.

good bet is to look at the first table of any research paper to see this information. Here you would see that the participants of this study with 6596 adolescents were 48.7% female, 70.9% white only,

74.2% aged 12–14 at the start of the study with more than one-third (34.3%) having parents with an education level of a bachelor's degree or more. Note that these are the categories of race/ethnicity and gender collected in this research. There are always other ways to measure variables. In terms of previous mental health status, the mean number of internalizing behaviors was 2.2 and externalizing behaviors was 3.2 with standard deviations of 1.6 and 2.1, respectively.

Social media time per day was grouped into five categories: none, 30 min or less, 30 min–3 h, 3–6 h or more than 6 h/day (i.e., a serious amount of social media use in a day!). In the total sample, about 17% spent no time on social media each day and about 8% spent more than 6 h/day. These summaries were also presented separately for the 611 participants who exhibited internalizing problems alone in Wave 3, for the 885 participants who exhibited externalizing problems alone in Wave 3 and for the 1160 participants who exhibited internalizing AND externalizing problems alone in Wave 3. Does this feel like many decisions are embedded in an analysis? The binning of social media time into categories with cut points of 30 min, 3 h and 6 h reflects an analysis decision. What if the cutpoints were 60 min, 4 h and 8 h? Would the analysis change?

Summary statistics are presented for the numeric variables and percentages for the categories variables.

> Descriptive statistics = summaries of the data in a sample.

It isn't helpful to have 6595 raw data values for the number of lifetime internalizing problems. If five or fewer different internalizing symptoms can be observed in a group of adolescents, then the data for the first 20 observations might look something like

1 0 1 0 2 2 5 4 1 0 0 0 2 1 0 4 2 2 3 5

with the first person in the sample reporting one symptom, the second no symptoms and the 20th person five symptoms. Not very interesting or insightful, particularly if you had a list of all 6595 values! A summary might be constructed such as a table with the percentage of observations at each value or some central value such as the mean number of symptoms. So what characteristics would you want to summarize? The frequency or percentage in each symptom category? The typical value of the symptom count? Variation in the symptom counts? All of these features are reasonable to consider.

To explore this further, suppose you generated 6595 random observations from a distribution with values that could be 0, 1, 2, 3, 4 or 5 with a mean of 2.19 and a standard deviation of 1.57, close to the values reported in the first table of the original research paper (Table 3.3). In this table, the relative frequency of symptoms showed that 26% or 1731 of the 6595 values reported one symptom. The most frequent score in a set of data is the **mode**, and for these data, the modal number of symptoms was 1.

Other common single measures are the **mean** or the **median**. The mean is the average number of symptoms observed, here 2.2. The median, here 2.0, is the score that splits the distribution of values into half with values less than the median and half greater. Thus, the median is the number of symptoms reported by the average individual as opposed to the mean which is the average number of symptoms in an individual (paraphrasing a D. Spiegelhalter description). Thus, the mean is the average number of symptoms observed in the sample; the median, on the other

TABLE 3.3   Frequency of Symptoms

No of Symptoms	% ($N = 6595$)
0	14.7
1	26.2
2	18.0
3	17.2
4	13.8
5	10.0

hand, is the number of symptoms reported by the average person in the sample.

Percentiles or quartiles are common summaries reported for describing the distribution of numeric variables. The median is the 50th percentile or second quartile (or 5th decile or ...). Other summary values of location such as the 25th percentile (aka first quartile) or the 75th percentile (aka third quartile) can be calculated. Here, 25% of people reported 1 or 0 symptoms and 75% of people reported three or fewer symptoms.

Now there is more to the story than a single summary value or average or typical scores. Consider three hypothetical samples of size 70 of a variable that can be recorded as 0, 1, 2, 3, 4 or 5 for each of the 70 sample members. The frequencies of variable values in each sample are summarized (Table 3.4).

While each of these samples has an average of 2.5, there are clearly very different patterns being suggested in these three samples. In Sample 1, 40 of the 70 responses are 2 or 3 and the scores <2 or >3 occur with decreasing frequency. In Sample 2, the most frequent scores are 0 or 5 with the least common responses being 2 or 3. Finally, the third sample has only responses of 2 or 3. Even with the different impressions suggested by the three samples, the mean and median for each sample is 2.5 symptoms! The missing part of the summary is the shape of the distribution of responses along with the variability in the responses.

TABLE 3.4   Frequencies of Variable Values in Different Samples with the Same Mean and Median

Score	Sample 1 Frequency	Sample 2 Frequency	Sample 3 Frequency
0	5	20	0
1	10	10	0
2	20	5	35
3	20	5	35
4	10	10	0
5	5	20	0

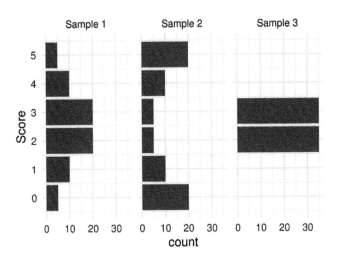

All three samples have the same mean and median but as the figure above suggests there is nuance to the story: the third sample has all participants with two or three symptoms, the second sample has most participants symptom-free (=0) and all symptoms present (=5) as the most common categories and the first sample has two or three symptoms as the most common but all other responses are possible.

Beyond the obvious shape differences, the range of the middle 50% of the data, the interquartile range (IQR) suggests that Sample 2 is more variable than Samples 1 and 3. Another common measure of variability,

> Be a critical reader:
> Don't tolerate someone reporting a summary of a central measure without some summary of variability. It is even better if you can see a display of the distribution of responses.

the **standard deviation (s)**, is related to the average deviation of scores around the mean. This measure suggests that Sample 2 exhibits the most variability ($s = 2.1$) followed by the first sample ($s = 1.3$) with Sample 3 exhibiting the least variability ($s = 0.5$) (Table 3.5).

What Is the Quality/Strength of the Evidence?

The summary statistics reported in Table 3.4 shows the percentage of social media use among all individuals (n = 6595),

> **Be a critical reader:**
> Always check to see the percentage of what group is being reported.

individuals with internalizing problems alone in Wave 3 (n = 611), externalizing problems alone in Wave 3 (n = 885) and both internalizing and externalizing problems (n = 1169). Here, 8% (571/6595) of all participants reported >6 h of social media use/day while 30% (171/1169) of participants with both internalizing and externalizing problems reported >6 h of social media use/day. This hints that the highest category of social media use is observed in adolescents who report both internalizing and externalizing problems compared to the percentage relative to the total sample.

Fitting a statistical model allows for a more formal evaluation of the impact of social media by adjusting other variables. The researchers

> **Reading Research:**
> Check out the second or third table in a research paper for model-fitting results.

used a model called multinomial logistic regression that examined whether the odds of being in one of the problem groups (internalizing problems only, externalizing problems only, both internalizing and externalizing) changed with increasing social

TABLE 3.5   Summary Statistics

Summary Statistic	Sample 1	Sample 2	Sample 3
Mean	2.5	2.5	2.5
Median	2.5	2.5	2.5
Standard deviation	1.3	2.1	0.5
25th percentile	2.0	0.0	2.0
75th percentile	3.0	5.0	3.0
IQR	1.0	5.0	1.0

media use relative to the odds of the problems in adolescents who did not use social media. The table of results in the original research paper included two columns of results for each problem group. Unadjusted analyses were presented along with adjusted analyses. In the **unadjusted** analysis, the impact of social media on mental health problems was investigated without considering any other variables that might impact mental health. In the **adjusted** analysis, the impact of social media accounting for or controlling for the impact of sex, race, age, parental education, BMI, alcohol use, marijuana use, and history of internalizing and externalizing problems on mental health problems was investigated.

> Reading Research:
> Adjusted analyses attempt to control for the impact of other potential confounding factors beyond the risk factor of primary interest (social media use here). Compare results from adjusted analyses to unadjusted analyses.

These models produced estimates that, when transformed, can be interpreted as odds ratios. The odds ratios of mental health problems were all calculated relative to the odds of problems in some groups. For example, the odds ratio of mental health problems for each social media use category was the odds of problems in that social media use category relative to the odds of problems in adolescents who did *not* use social media. In addition to an estimate of the odds ratio, an interval estimate was reported. The **point estimate** is the single best estimate of some characteristic (population odds ratio here), and the **interval estimate** provides a plausible interval of values that reflect some of the uncertainty in estimating population features.

> Be a critical reader:
> A false sense of precision is often conveyed when only a single number is presented. Expect interval estimates or margin of errors in summaries and stories based on research.

Let's look at one of the values reported. The odds value for both internalizing and externalizing problems in adolescents using social media for 3–6 h each day was 2.01 times the odds value for both problems among adolescents who did not use social media, i.e. OR = 2.01 with a 95% confidence interval of 1.51–2.66. In addition, the estimated odds ratios for both problems increased with increasing social media use which is supporting evidence that a relationship between a risk factor and a response might be causal.

It is worth noting that the adjusted OR values were all smaller than the unadjusted OR values. Compared to the adjusted OR = 2.01, the unadjusted OR = 3.15, which suggests that the relationship of problems with social media use might be confounded by other factors. In this study, males are estimated to have half the odds of females for both problems (CI: 0.43–0.61) and adolescents aged 15–17 are estimated to have odds of 0.82, the odds of adolescents aged 12–14 (CI: 0.70–0.96).

IS A 2X INCREASE IN ODDS OF PROBLEMS A CAUSE FOR CONCERN?

Often the answer to a question like this is "maybe." If 100 youths with social media exposure of over 3 h/day have 20 mental health issues compared to only ten mental health issues observed in 100 youths with no social media exposure, this might be of grave concern. In contrast, if 10,000 youths with social media exposure of over 3 h/day have two youths with mental health issues compared to only one mental health issue observed in 10,000 youths with no social media exposure, this might not be a story worth noting. In both scenarios, there is a doubling of the relative risk; however, in terms of absolute impact, the first scenario reflects a more serious situation. Relative measures are often headline-grabbing and a tasty part of any story about the

> Be a critical reader:
> Relative risk measures (ratios) are often reported but absolute measures (differences) are a better reflection of impact.

effect of some exposure; however, they are not meaningful unless you have an idea about how often you would observe the response in the absence of the exposure.

WHAT ARE THE BASELINE RATES OF THESE MENTAL HEALTH PROBLEMS?

This can be difficult to evaluate, particularly when the baseline rates of mental health problems may differ between covariates such as males and females, age, BMI and other covariates. In the research article, the probabilities of mental health problems were predicted using the statistical model with all of the covariates set at their mean values. In this case, about 10% (~8–12%) of adolescents who do not use social media are predicted to experience both internalizing and externalizing behaviors, and about 17.5% (~16–24%) of adolescents who use 3–6 h of social media each day are predicted to experience this behavior.

IS THE CLAIM REASONABLE IN ITSELF? DOES PRIOR BELIEF IMPACT MY BELIEF? CONFIRMATION BIAS?

The study took care to look at social media use in a time period that preceded the time period when mental health was evaluated. A large set of predictor variables, beyond the social media use risk factor, were considered, and these included some history of mental health. We are not sure that we had a strong prior belief about the relationship between social media use and mental health status. If pressed, we might have predicted that mental health problems would increase with increasing levels of social media use; however, we also might have predicted the opposite assuming that social isolation as indicated by no social media use might have been more indicative of mental health problems.

HOW DOES THIS CLAIM FIT WITH WHAT IS ALREADY KNOWN?

The introduction and discussion in a scientific paper are where the motivation for research is described relative to the existing literature and where the results are compared to other evidence in the research literature. These scientists cite studies, both cross-sectional and longitudinal in nature, where social media is related to internalizing problems although they caution that this is not a consistent result. The authors situate their own research as building on the literature about the relationship between adolescent mental health and social media use.

When discussing this research, the authors note the consistency of the observed result related to social media impact with previous work. A study that observes effects that are similar to those observed by other researchers or conducted in other populations adds support to a particular theory under consideration.

> Reading Research:
> A good introduction will summarize other research (filtered through the perspective of the authors).
> A good discussion will place results in the context of other work and highlight limitations of the work along with future directions.

Researchers also note where their results differed from their expectations or from the work of other researchers, what were some of the limitations of the work and what they see as the next research question to examine. These researchers mentioned the potential problems with self-reported social media use and mental health determined from a short questionnaire along with other potential concerns. A question that almost always can be raised is "are there other variables that might explain both decreases in mental health AND increases in social media use?" These are sometimes called **lurking variables** (does this bring up an image of a burglar about to sneak into a house?).

HOW MUCH DOES THIS MATTER TO ME?

As residents of the United States, a study of U.S. teens is targeting the youth in our country. Would we recommend restricting social media use among adolescents as a consequence of this work? Maybe a couple of years ago, but we're not sure we would now. What is the relevance of data from a 2013–2016 study during times of social distancing and pandemic? (Note: We finished our work on this book as things in the United States began to open up in the summer of 2021.) It may be that social media use now is a predictor of healthy connection versus a predictor of mental health problems when face-to-face social interactions were possible. It's always worth considering whether the conditions when a study was conducted still apply. Our experiences during the COVID-19 pandemic suggest maybe not in this case.

DOES A STUDY OF U.S. YOUNG TEENS TRANSLATE TO OLDER TEENS OR TO OTHER COUNTRIES?

This was a study of U.S. 12- to 15-year-olds. If this was a representative sample of the population of 12- to 15-year-olds in the United States, then estimates and conclusions from this study would be relevant for this population. There would not be a debate about whether this study can be used to draw conclusions about 12- to 15-year-old U.S. kids. You might wonder if social media impacts 12- to 15-year-old kids living in cities in the same way as 12- to 15-year-old kids living in the suburbs or country. It may be that sufficient information is available on subgroups but often it is not.

More generally, would you expect conclusions drawn from 12- to 15-year-old U.S. adolescents to apply to 10- to 12-year-old (what are they doing on social media for 3 h?) or to 16- to 18-year-old youths/young adults? This is a similar cultural and national context; however, the generalization of the 12- to 15-year-old sample to populations with different ages is another example of an **extrapolation**. It might be that the different age groups have the same relationship between social media use and mental health, but there is nothing in this study that would support this. Another extrapolation might be to consider if this social media–mental

> Be a critical reader: Check to see that the population being studied and impacted by some factor is the population of interest to you.

health relationship would be the same in youths of the same age but who live in other countries. Why would the relationship be the same in Taiwanese 12- to 15-year-old youth as in U.S. 12- to 15-year-old youth? Our suspicion is that this would be a stretch.

CONSIDERING THE COVERAGE

The article reporting on this study is really more of what we'd call a news summary or news stub in the world of journalism. A full-fledged article would feature quotes and dig a bit more in-depth into the study's conceptualization and findings. This piece from the *Washington Post* is really just a summary of the study's findings without any real deep engagement with much in the study. What might a fully reported news article have done differently? Well, for starters, it would have worked to better contextualize the findings.

> A news stub is a short article that gives an audience a rundown of the basic facts of a story, without deeply interrogating or contextualizing them.

As part of the reporting process, journalists are tasked with searching for alternative explanations for claims made by the sources they interview – whether those sources are politicians, scientists or soccer fans. In fact, the Society of Professional Journalists' Code of Ethics urges reporters to "Provide context. Take special care not to misrepresent or oversimplify in promoting, previewing or summarizing a story" (Society of Professional Journalists, nd.). While news summaries have their utility, reporting on studies in such summaries or blurbs does not allow for a robust interrogation of facts or claims. This is especially problematic if the news outlet never follows up that early summary story with a more fully realized piece of reporting. For some issues, this may not be a tragedy; however, in the case of stories about the effects of media use on mental health, this is misleading for many reasons.

First, communication research has shown over and over again that media do not generally have direct effects on audiences. In the early 20th century, proponents of the hypodermic needle theory of communication imagined that whatever message was transmitted in media would shoot through public conscious-ness, inoculating people against certain ideas. (This theory grew out of studies of propaganda – produced both by people who were concerned about its effects and those who hoped to harness propaganda's potential power to sway public opinion.) To sum up decades of research in a sentence: scholars have since shown that our relationship with and reception of media are terribly complicated – shaped by politics, history, interpersonal relationships and the cultures in which we live. We do not swal-low, wholesale, the narratives pushed in media. When it comes to social media, what we get from our experiences in them is dependent on what we bring to them, who we connect to as well as the realities of our lives outside social media. It is hard to prove causation when it comes to the impact of media use – whether we're talking old or new media. A news article that did more than summarize findings might have begun to dig into this. After all, the study's authors note that social media have also been predicted to have positive mental health outcomes, something the *Washington Post* news stub doesn't really discuss.

It is worth considering how new technologies may secure addi-tional attention in reporting. Here, the other issue with reporting on social media is the weight that is sometimes given to new media technologies – it is a weight that often takes away the agency of the individuals who use them. For instance, in a study of the news coverage of the Steubenville rape case, Rosemary Pennington and Jessica Birthisel (2016) found that journalists often covered the story in a way that made it seem like the new media technolo-gies involved in the case (cellphones and Twitter, mainly) were to blame for the assault – not the young men who were ultimately convicted of the rape. Such reporting can make it seem like users have no control over how media affect them, although the *Washington Post* news stub does note that the researchers [in the John Hopkins study] suspect that heavy use may lead to sleep problems that can contribute to such issues, increase the risk for cyberbullying, which has been tied to symptoms of depression, and result in unrealistic comparisons of yourself and your life to those of others seen on social media.

This might feel like overreach or an extrapolation well beyond what the study suggested. However, it's also a reminder that a single study is part of a continuing investigation into phenomena. A study often suggests directions for future investigations that could resolve why something happens but rarely, if ever, completes the story as to why something happened. Other reporting on the Johns Hopkins study did attempt to contextualize its findings, even if the headlines seemed to sensationalize the study.

An article in the *MIT Technology Review* titled "Teens are anxious and depressed after three hours of social media"[1] walked the reader through the highlights of the study and included a section called "But this is old news... right?" which put the Johns Hopkins study in conversation with other research on the topic, including a study which suggested that no link between the use of social media and mental health exists. An article in the *Philly Voice* took a more modulated tone in its reporting of the study's findings, starting with its headline which told readers the study found that social media "hikes risk of mental health issues for teens."[2] In its first sentence, the story reiterated that the study found teens at "higher risk," which is a qualification that seemed to be absent from other reporting on the study. It also provided a bit more detail about how the study was conducted and included a quote from the study as well. It did not provide the kind of contextualization found in the *MIT Technology Review* article, but it did give readers a bit better understanding of how the study was done than the *Washington Post's* news stub provided.

The Johns Hopkins study is certainly a newsworthy event, as many of us are concerned about the impact of social media on our lives. (Since we worked on this chapter, information has come out that Facebook was aware that its app, Instagram, makes body image worse in its teenage female users.) Short news summaries or news blurbs have their place in our news consumption diet – not every story can be an 8-min long public radio feature – but they do not provide much of a venue for journalists to explore the broader implications of a study nor consider, deeply, other explanations or interpretations of a study's findings. Stories like the *Washington Post* article we discussed in this chapter are a good starting point for learning about a study, but if we hope to better understand the context of a piece of research, we should either read the study itself or seek out longer news stories that do more work contextualizing the research or considering other interpretations.

REVIEW

A survey of U.S. adolescents was conducted and measurements were taken over three years. Even after adjusting for other features that might impact mental health, self-reported mental health problems are related to self-reported social media use. The research summarized in the story was careful in study design to collect mental health measurements after social media usage and to control for confounding factors. Both relative and absolute risk estimates were provided. However, if an individual only read the *Washington Post* article about the study, they might get the mistaken idea that the science on the relationship between social media use and teen mental health is more settled than it really is.

STATS + STORIES PODCASTS

Stats + Stories featured a conversation with science writer Christie Aschwanden about reporting about science and health (https://statsandstories.net/health1/getting-health-and-science-reporting-right).

Ty Tashiro discussed social awkwardness and relationships in two episodes: https://statsandstories.net/lifestyle1/2018/8/2/reading-the-book-of-love-what-can-you-learn-from-relationship-

science-stats-stories-episode-50 and https://statsandstories.net
/lifestyle1/2018/8/2/chins-and-ears-are-not-information-rich
-awkwardness-and-social-relationships-stats-stories-episode-57.

Jessica Myrick commented on emotional responses to social
media in https://statsandstories.net/lifestyle1/2018/8/2/stats-of
-cool-cats-emotions-mood-management-and-cat-videos-stats
-stories-episode-41.

NOTES

1. Teens are anxious and depressed after three hours of social media,
MIT Technology Review. Accessed at https://www.technologyre-
view.com/2019/09/11/133096/teens-are-anxious-and-depressed
-after-three-hours-a-day-on-social-media/
2. 3 hours a day on social media hikes risk of mental health issues for
teens, study finds, Philly Voice. Accessed at: https://www.philly-
voice.com/3-hours-day-social-media-hikes-risk-mental-health
-issues-teens-study-finds/

CHAPTER 4

Speedy Sneakers

Technological Boosterism or Sound Science?

Researchers say a new Nike shoe can actually make you a faster runner – November 20, 2017. Story Source: Quartz (by Marc Bain) – https://qz.com/quartzy/1133811/nikes-zoom-vaporfly-4-makes-you-a-faster-runner-study-says/

Scientific Journal Citation: Hoogkamer, W., Kipp, S., Frank, J.H. *et al.* A Comparison of the Energetic Cost of Running in Marathon Racing Shoes. *Sports Med* 48, 1009–1019 (2018). https://doi.org/10.1007/s40279-017-0811-2

https://link.springer.com/article/10.1007/s40279-017-0811-2

DOI: 10.1201/9781003023401-4

(Photo by RUN 4 FFWPU from Pexels) (https://www.pexels.com/photo/female-and-male-runners-on-a-marathon-2402777/)

STORY SUMMARY

Athletes, amateur and professional alike, are constantly looking for things that can help them be better, stronger and faster. This story by Marc Bain writing for *Quartz* focuses on a new running shoe by Nike. The shoe, the Zoom Vaporfly 4%, is supposed to help runners run faster using 4% less energy than they would normally (or wearing other shoes). In his reporting, Bain details a study by University of Colorado researchers that seems to suggest the shoe might actually live up to its promise. The story features quotes from the study that detail just how the researchers went about testing how much zoom the shoes possess. The reporter then details what real-world evidence might be necessary to prove the shoes' benefits hold up outside a lab setting – it's one thing to be faster in a controlled environment, another to be faster when contending with bad road or weather conditions. The story wraps up with a consideration of what this kind of technology might mean for the future of running as a sport.

WHAT IDEAS WILL YOU ENCOUNTER IN THIS CHAPTER?

- Participants in experiments may be selected based on certain criteria – consider how these study subjects might differ from you.

- The variable that is measured may be a surrogate for the true response of interest.

- Good experiments control for extraneous sources of variability that might impact a relationship of interest.

- Well-designed experimental studies share common key ingredients.

- There is logic used when testing hypotheses about treatments or different factors (including types of running shoes!).

- A news story is first framed for a particular focus.

- Uncritical journalistic boosterism is a factor to consider when examining stories.

WHAT IS CLAIMED? IS IT APPROPRIATE?

The headline for this story claims that a new shoe makes you faster. This featured sports story from *Quartz* summarized a scientific paper published in the journal *Sports Medicine*. However, before you go out and buy new kicks, the story notes that this new shoe was designed for "serious marathoners" and that it used 4% less energy, on average, than the previous top racing shoe manufactured by Nike. The chain of logic is that less energy translates into an ability to maintain a faster pace and thus a faster time for a race.

WHO IS CLAIMING THIS?

Researchers at the University of Colorado are claiming that the energy expenditures associated with running in a new shoe were less than running in shoes from other companies. This university-based research was funded by the shoe manufacturer Nike and included two authors who were employees of Nike. The story notes that researchers disclosed the conflict of interest and received ethical approval to conduct the study from the local university institutional review board. Research requires funding. Knowing who funded the research may help you evaluate whether a published result is what you might expect.

> Reading Research:
> Always ask if the authors have a vested interest in the study outcome.

> Reading Research:
> Research requires approvals, particularly if human subjects are involved. Institutional review boards evaluate whether the merit of conducting a research study offsets the risk for participants.

WHY IS IT CLAIMED?

An experiment was conducted to evaluate the "energetic costs of running" with a newly developed running shoe versus two shoes routinely used by marathoners – another Nike shoe and the Adidas shoe worn by the then-world-recorder holder. The study authors noted that these two comparison shoes (or previous versions of these shoe designs) were worn by the 10 fastest competitors to date. The researchers conducted an experiment on the impact of shoe type on energy use. As we see below, experiments typically include some aspects of randomization, replication and control comparisons.

The study covered by Quartz actually reported two experiments: (1) machine-induced physical stress that produced measurements of "mechanical energy stored and returned" for each shoe; and (2) a study of elite runners using these shoes on a treadmill.

Photo by William Choquette from Pexels (https://www.pexels.com/photo/an-on-treadmill-1954524/)

The machine mimicked what happens to a shoe when you run – your foot hits the pavement; the shoe cushions your step by the sole compressing and then the sole of the shoe returns back to shape. The new shoe design had a greater "energy return" than the two other shoes.

The experiment with

So why combine the machine study with the human experiment?

The mechanical response might not translate directly into human performance. The human study adds "ecological" validity to the analysis since the comparison is in runners using the shoes in (laboratory) natural conditions.

the runners measured energy consumption, "ground reaction forces" and lactic acid build-up when running a 5-min test run. Each runner ran in each of the three shoe designs at three different training speeds (14, 16 and 18 km/h). Runner physiological and "energetic responses" were measured. The energy use by runners wearing the new shoe was reported to be about 4% less on average than the energy use by runners wearing the two comparison shoes.

IS THIS A GOOD MEASURE OF IMPACT?

It is hard to imagine evaluating different shoe designs using any measurements better than the experience of runners using these shoes. The researchers noted that running speed is determined by oxygen consumption, lactate thresholds and "energetic cost of running," with the last feature being of primary interest to the researchers. One serious question to consider is: Are you a member of the same population from which this sample of runners was obtained?

Reading Research:
Don't expect to know about all the variables in a study unless you happen to be a specialist in this area. Lactate threshold anyone?

HOW IS THE CLAIM SUPPORTED?

The researchers described the design of the new prototype shoes and properties of the shoes, talked about a mechanical study to test these properties in the three shoe brands being evaluated and reported a study of runners designed to explore human experience with the shoes. The study of the runners involved the comparison of the three shoe types at three different running speeds that were set on a treadmill.

What Evidence Is Reported?

How would you design an experiment to evaluate the performance of different running shoes? Would you recruit volunteers from your local running club to run in these shoes and report back to you? Perhaps you

Reading Research:
Look for details about randomization, replication and control when reviewing experiments.

might ask one-third of the volunteers to use the prototype shoe, one-third to use one of the current brands and one-third the other current brand. Not a bad idea but what if you are unlucky and all of the best runners in the club were assigned one of the shoe types? Ideally, the groups should be as similar as possible prior to evaluating the differences between the shoes. As we see below, each participating runner ran in each shoe type; a comparison of shoe type controls for differences that might exist between each runner. Studies of experimental factors should exhibit some randomization of conditions, some replication of these conditions and some control or comparison group.

Experimenters often impose selection criteria on participants in a study. At a simple level, this filters out some of the variability that might be expected in study participants. A set of male, 31-min/10 km, size 10 shoe runners is clearly a subgroup from the population of all runners. This was a high-performing group who could successfully complete a study where they ran six 5-min "trials" during a day.

Randomization: As hinted above, the simplest type of randomization in an experiment would randomly assign shoe styles to some subset of runners in a study. In such studies, a set of runners might be split into three groups where each group would be assigned a particular shoe type to use for measuring responses of interest. This type of design would use some randomizing mechanism (think the equivalent of observing the result of tossing a three-sided coin – not very useful for slot machines but might work as a mental image) to do the shoe-type assignment in hopes that the runners in each of the three groups would be as similar as possible prior to running in the new shoes. Generally, such an experimental design would require more runners in the study than any alternative design such as the one used here.

In this study, each runner did a 5-min treadmill run at a specific pace in each shoe (with a different pace set on three separate days of assessment), running at that pace in each shoe twice a day. Thus, a runner ran six 5-min runs each day, and each runner was randomly assigned to a sequence of runs with shoe types each day.

A "mirrored order"/ sequence of runs on a day with three runners randomized to each of six sequences. If the shoe types are labeled S1, S2 and S3, then the possible mirrored orders would be

S1-S2-S3-S3-S2-S1

S1-S3-S2-S2-S3-S1

S2-S1-S3-S3-S1-S2

S2-S3-S1-S1-S3-S2

S3-S1-S2-S2-S1-S3

S3-S2-S1-S1-S2-S3

Here, the first sequence, S1-S2-S3-S3-S2-S1, corresponds to running the first 5-min run using shoe brand S1, the second 5-min run using shoe brand S2 and so on until the final 5-min run was using shoe brand S1 again. Why do these sequences instead of S1-S1-S2-S2-S3-S3 or some other pattern? What if there was a systematic difference in energy use in the first of the 5-min runs versus the last runs? You wouldn't want this to confound the comparison of the shoe types. This randomization of a sequence of runs helps avoid this confounding. Other key ideas to keep in mind include the following:

Replication: No one would believe a study with a single runner in each shoe type. Having more than one runner allows you to explore the impact of shoe type in the presence of the variation you would expect among runners. The question that is often part of good statistical thinking is: Do you observe a signal amidst the noise of different subjects?

Control: Experiments usually will compare some new product/ treatment to existing products/treatments. The new prototype shoes in this particular research were not compared to running barefoot, they were compared to the shoes worn by the best, most elite runners in recent races. It is news if the new prototype is better than the *best* shoes on the market at the time, but not very exciting if the new prototype was only superior to running in flip-flops or on bare feet. Another way to think of this is the "control" condition could be a standard condition or current best practice. The use of a no-treatment control group in a study of a new medical treatment for some diseases would be unethical if effective treatments were already available.

What Is the Quality/Strength of the Evidence?

Each of the two runs per shoe type was averaged for each runner for this analysis. The analysis involved the comparison of two factors – shoe type (primary factor of interest) and running speed. Each runner contributed 18 measurements to the running part of the study: 3 running speeds × 3 shoe types × 2 replicates/running speed–shoe type combination. They also contributed additional information useful for baseline performance and other responses.

The researchers did this study because they believed that the new prototype shoe was designed to be superior to existing shoe technology. This is as analogous to a new pharmaceutical product being tested because it is believed to be superior to existing products. However, what they might find is that the new prototype shoe is no different than the other shoe types being considered. This is where data from an experiment come into play. If you were going to conduct research to measure the performance of something – in this case a running shoe prototype – what might you do given what you've read so far?

The simple response to the question above is to collect running performance data for the three shoe types and see if the new prototype uses less energy than the comparison shoes. Here is where statistical thinking is needed. You know that even if the shoes were the same in terms of energy use, you may see the prototype using less energy by chance.

Let's simplify the experiment to explore this in more detail. Suppose you want to compare only two shoe types. If you compared the experience of 18 runners and observed the

> Reading Research:
> Formal hypothesis testing involves looking at sample data to decide between two competing hypotheses about a population.

prototype shoe being superior to the other shoe nine times, you might say, "This is what I might expect if the new shoe was similar to the old shoe." Ok. So what if 18 runners resulted in the new shoe judged superior by 10 runners? Here (channeling you again),

> It is not surprising to see 10 runners with the new shoe exhibiting superior performance even if the shoes were similar. In a more trivial coin-tossing-curse-of-introduc tory-statistics-course analogy – you would not be surprised that a fair coin tossed 18 times results in 10 heads even though you expected 9.

Ok, how about if the prototype-old shoe test with 18 runners results in 11 or 12 or 13 or 14 or 15 or 16 or 17 or 18 times with the prototype preferred? Would you still say, "I believe the shoes are the same in terms of performance" when you observe 18 runners with the prototype preferred OR would you reject a belief that these shoes were comparable and conclude the prototype shoes were superior? How much evidence would you require to declare the prototype shoe appears superior? Here the testing process for this simple study might be summarized as:

Research hypothesis: Prototype shoe superior to the other shoe type.

No-effect hypothesis: Two shoe types are the same in performance.

Data collected: Number of times the prototype shoe was preferred in a test of 18 runners.

How much data is needed for you to conclude the prototype shoe is superior?

So, how do you decide how "superior" is defined? In this study, superior is defined in terms of the mean energy expended by a population of runners using the new prototype shoe being less than the mean energy expended by a population of runners using the current shoes.

Could you make a mistake? Sure. You might conclude the prototype is superior when it's not or conclude the shoes were similar when the prototype was superior. In formal testing of hypotheses, investigators will often set

> Errors can occur in decision-making: You can decide there is a difference when there really isn't (aka **false positive error**) OR you can decide there is no difference when there is really a difference (aka **false negative error**).

an acceptable level of one of these errors – here, the researchers reported that a **level of significance** of 5% was used to declare that a difference in energy use exists among the shoes. In this case, if there was truly no difference among the shoes, then there is only a 1 in 20 chance (5%) that the study would declare a difference exists.

Many research reports will summarize statistical tests using **P values**; quantities that are often compared to these levels of significance. A P value represents the unusualness of an experimental result assuming a particular hypothesis is true and assuming some underlying statistical model is true. Small P-values (close to zero) suggest that something unusual has occurred, including the hypothesis you assumed is not correct or the model underlying the calculation is not correct. The use and misinterpretation of P-values led the American Statistical Association to publish a statement on P-values and significance testing in 2016 (https://www.amstat.org/asa/files/pdfs/P-ValueStatement.pdf) that should be required reading for background use and consumption of P-values.

IS THE CLAIM REASONABLE IN ITSELF? DOES PRIOR BELIEF IMPACT MY BELIEF? CONFIRMATION BIAS?

Our prior belief was that the technology associated with sports performance would improve over time. Thus, you may not be surprised by the result from analyzing the human runners or from the machine tests of the shoes.

The experimenters essentially used each runner as their own control since every runner ran in each of the three shoes. As such, the differences in the shoe brands are evaluated within the same runner, and then the results are pooled over runners for the analysis. The experimenters also controlled other potential confounding factors to the extent possible by imposing selection criteria for participants (you had to be a serious male runner to qualify for the study) and adding weights to shoes so they would be as similar as possible prior to the run (one brand was heavier than the others so this adjustment made the shoes comparable with respect to this important determinant of energy use). Ultimately, the experiment merited the conclusion in the headline that the new shoe used less energy and the implication that this would translate into faster running times seems like a reasonable stretch.

HOW DOES THIS CLAIM FIT WITH WHAT IS ALREADY KNOWN?

The researchers cited literature that identified factors that impact running velocity (oxygen uptake, lactate threshold and energy cost). By including elite runners they essentially had study participants with similar oxygen uptake and lactate thresholds. This allowed them to compare energy

Reading Research
The Introduction (first section) of most articles presents the context and motivation for research studies.

The Discussion (last section) puts the experimental results in context along with identifying limitations, caveats and potential future work.

costs directly. The authors cite the literature related to running shoe design and energy cost. The research study wraps up with citations concerning the relationship of shoe design to the energy costs of running and explanations of why this might be expected. Predictions of how a 4% energy savings might translate into a world record marathon time were also discussed, although the researchers noted that the fastest pace studied in the experiment was slower than the pace set in the world record marathon time.

The *Quartz* story included an interview with independent experts who reviewed the results. One expert affirmed that the study methods were appropriate for studying the energetic costs of running in different shoes, but would have felt more confident in the conclusions with larger sample size. The other expert suggested that runners might run differently if they know they are using some new prototype shoe. The idea that your response might be impacted, possibly subconsciously, by knowledge of the treatment (shoe) you received (are running in) is another factor that impacts a result. This is a reason why medical studies use blinding in study design (e.g. a double-blind study is one where neither patient nor physician knows the treatment being received) and why experiments look to control for potential placebo effects in studies.

HOW MUCH DOES THIS MATTER TO ME?

This study evaluated shoes in 18 male athletes who could run a 10 km race at a 3.1 min/km pace (aka a 5-min mile pace for those of you morally opposed to metric measures) or comparable pace. In addition to being really fast runners, these athletes needed to fit into size 10 shoes.

The only qualification that one of us could meet for participating in this study is that John can wear size 10 shoes. Even on his most fit running days, John could not have qualified for the pacing requirement. In fact, he confesses that couldn't have achieved a 5 min/mile pace even if being chased down a gentle hill by the Slenderman, a mountain lion or just a very fat dog.

The issue then becomes how do we think about this result? Clearly, these runners are sampled from a population of which one of us is a member; however, they are from a population of elite runners and, thus, this result suggests that new running shoe technology may translate into new records. The results of this study are still intriguing for more recreational runners. Less energy cost should translate into greater velocity regardless of the quality of a runner. If more energy is returned based on the responses in the mechanical study and less energy is used by elite runners, then a recreational runner might want to see if there was a benefit. Finally, this study begs the question of whether a similar shoe is available for women runners and, if so, would the results be expected to be the same?

CONSIDERING THE COVERAGE

When a journalist sits down to write a story, one of the first things they have to decide is what angle to take with it. Every choice from that moment onward influences how the journalist frames the story they are telling. The concept of framing is an important one in the field of communication and media studies – and the one we introduced in Chapter 2 – but it's also a concept that has a number of different definitions. Perhaps the most widely cited definition is that of Robert Entman. For Entman (1993)

> To frame is to *select some aspects of a perceived reality and make them more salient in a communicating text, in such a way as to promote a particular problem definition, causal interpretation, moral evaluation and/or treatment recommendation* for the item described (italics in copy).
>
> (P. 52)

The framing of a news story shapes how we understand and interpret an issue or situation. Sociologist Erving Goffman pointed out that frames are socially and culturally constructed; they live in our heads and all of us, journalists or not, rely on them to help make sense of our world. Deciding to frame a news story in one way forces a particular interpretation of an issue.

The *Quartz* article is an interesting case in news framing. On its surface, it's a science story, but the author is described as a "fashion reporter." What seems to emerge from the mixing of these two things

> **Journalistic boosterism** is reporting that focuses on the good or positive aspects of something, often in order to promote it or to boost its visibility. It is most often associated with the coverage of politics – at both the local and national levels – but can also be found in the coverage of such things as sports teams or even scientific discoveries.

– science and fashion – is a tension between verification of the claims and fascination with the new design. One of the things reporters covering new technologies – fashion or otherwise – have to be careful of is falling into an "isn't this amazing" framing. Not only does it suggest a kind of boosterism that is inappropriate for news coverage, but it can also lend itself to an uncritical recitation of claims. Boosterism often leads to reporters not as fully verifying the facts of a story or vetting sources as thoroughly as they might. This tension between verification and fascination can be found in the very headline itself, "Researchers say a new Nike shoe can actually help you run faster" under its subhead of "Zoom!"

What makes it even more problematic is that readers don't find out that Nike funded the study or that Nike researchers were involved in it until the fourth paragraph. Using the word "researchers" in the headline makes it seem like these were independent scholars who conducted the work. A better headline might be, "Nike claims a new shoe helps you run faster. But does it?" It might not have the zing of the original, but it does a better job of qualifying the study's claims and making clearer Nike's own boosterism of this shoe, while also putting some distance between the reporter and what they are reporting on. The reporter does seem to work away from this boosterism a bit as he moves into a discussion of the outside experts' views on Nike's study, but even here is a bit of a reporting problem as for the first-time readers learn that the runners in Nike's experiments knew they were using a new shoe. The reporter notes "The flaw they identified was that the test runners knew they were running in a new shoe from Nike, which could have created a placebo effect … a placebo effect on its own can improve performance."

So, we have two not insignificant details – the involvement of Nike in the research and the possible placebo effect – buried deep in the story. Journalists are supposed to approach their reporting with a critical eye – journalism professors often tell their students they should imagine their role as that of professional skeptic – and the reporting in this *Quartz* article is not as critical as it might have been. Again, this is not about being critical in a negative sense, but simply a reminder that journalists are charged with seeking out alternate interpretations for a claim before framing it as the truth. It's really not until the very end of the story that the reporter seems to apply that kind of critical lens to his reporting. Again, that tension between verification and fascination we mentioned earlier seems to have played out through the whole article, which is exemplified in these final sentences of the story.

> We also can't say for sure how these findings translate outside of lab conditions. Runners have been wearing the Zoom Vaporfly, but their results, while noteworthy, haven't been abnormally fast. Nobody has completed that sub-two-hour marathon just yet.
>
> It could be only a matter of time, though. In May, Nike had some of the world's best distance runners complete a marathon in custom Vaporfly shoes. Eliud Kipchoge, the world's best marathoner, finished in two hours and twenty-five seconds.

The fascination, however, might have been well placed. A number of articles have been published since this 2017 study first came out suggesting that Nike's shoes actually do help runners run faster. A 2020 article from *Runner's World* claims "They are unequivocally the fastest shoes money can buy," so fast, in fact, that they were almost banned from international competition (Roe 2017). The reporter then goes on to detail how the folks at *Runner's World* dissected the Nike shoe, and similar shoes from two competitors, to understand what makes it so fast. (Eliud Kipchoge, by the way, did break the two-hour marathon barrier – thought to be an almost impossible feat – with the *Runner's World* article suggesting he may have been wearing a prototype of a new version of the Nike Vaporfly.)

Does the apparent speediness of Nike's expensive running shoe justify the early framing in the *Quartz* article? Not necessarily. The piece in *Runner's World* was written after a number of marathoners had worn the shoes while clocking in impressive times – providing real-world evidence in support of those initial lab findings. But, the *Quartz* reporter could not have predicted that. Right now we're talking about research about a shoe which, on its face, is pretty low stakes. But think about reporting on research on different types of scientific discoveries – treatments for things like cancer or the development of vaccines for things like COVID – would you be comfortable reading reporting that veered into boosterism territory in those cases? The way a story is framed influences how the public understands something – it can make it appear the scientific consensus around something is more settled than it actually is and vice versa. As you read news stories about experimental research, you should ask yourself why the story is framed the way it is and whether the reporter has done enough to vet the claims in the story.

REVIEW

This story reported the results of a new prototype shoe that was designed to use less energy when running. A study of runners was conducted to compare the prototype with two leading shoe types used by elite runners. The study involved the principles for good experimental design including the randomization of runners to a shoe-use sequence, replication of each sequence, with runners serving as their own control for comparisons between shoes. A formal comparison between the mean energy costs between the shoes was made, and a 4% difference between the prototype and the two other shoes was observed which was formally assessed using statistical hypothesis tests. There is logic used when testing hypotheses about treatments or different factors (including types of running shoes). However, the reporting on the study's findings was less critical than it probably should have been, leading to a kind of boosterism that does more to support a study's claims than actually vet them.

TO LEARN MORE

Check out statistical textbooks to learn more about the statistical hypothesis testing methods described in this study. The statistical model used in this analysis had to account for the fact that multiple observations were taken for each runner.

A BONUS STORY

> Are elite runners getting faster, or is a special pair of Nike shoes giving them an edge? – January 29, 2020
>
> Citation Story Source: https://www.nbcnews.com/ business/consumer/are-elite-runners-getting-faster-or-special-pair-nike-shoes-n1121921

An *NBC News* follow-up story was released a little more than two years after the original story about these new Nike shoes. These prototype shoes in 2017 are now the source of debate by sports officials who are investigating the availability of these shoes for runners. In six marathons run in 2019, this story reports that more than 85% of the 36 top finishers ran in these new shoes. A sub-2-h marathon was run in October 2019 (https://olympics .nbcsports.com/2019/10/12/eliud-kipchoge-marathon-two-hours

-ineos-159-challenge/); however, this was a marathon run under ideal conditions for this racer who was accompanied by other runners who helped pace this race, and thus this time was not eligible for world record status. Ultimately, the 2017 story with its prediction of faster races and the breaking of the sub-2-h marathon was realized.

STATS + STORIES PODCASTS

If you're interested in exploring other stories about stats in sports, the Stats + Stories podcast has featured a number of interviews on the topic (https://statsandstories.net/sports).

Author Jim Albert talks about sabermetrics and player and team management in the episode "Baseball and Statistics" from 2013 (https://statsandstories.net/sports1/2018/8/2/baseball-and-statistics-stats-stories-episode-1).

Sports reporting was the focus of an episode with guest Terence Moore (https://statsandstories.net/sports1/2018/8/2/sports-reporting-in-the-digital-era-stats-stories-episode-18).

Football was the focus of a conversation with journalist Alan Schwarz as he discussed the prevalence of brain injuries in American football (https://statsandstories.net/sports1/2018/8/2/gridiron-touchdown-field-goal-traumatic-brain-injury-stats-stories-episode-39).

Analytics for football was the conversation with Dennis Lock (https://statsandstories.net/sports1/2018/8/2/a-winning-formula-for-sports-stats-stories-episode-21).

The sport the rest of the world knows as football, but Americans as soccer, was the focus of a conversation with Luke Bornn (https://statsandstories.net/sports1/2018/8/2/g-o-o-a-a-a-l-l-l-l-celebrating-the-statistics-of-the-beautiful-game-stats-stories-episode-59).

CHAPTER 5

Investigating Series Binge-Watching

What Happens to your brain when you binge-watch a TV series – November 4, 2017 (by Danielle Page, NBC News). Story Source: https://www.nbcnews.com/better/health/what-happens-your-brain-when-you-binge-watch-tv-series-ncna816991

Netflix Media Center Report: Netflix & Binge: New Binge Scale Reveals TV Series We Devour and Those We Savor.

https://media.netflix.com/en/press-releases/
netflix-binge-new-binge-scale-reveals-tv
-series-we-devour-and-those-we-savor-1

Government Report Citation: American Time Use Survey Summary

https://www.bls.gov/news.release/atus.nr0.htm

DOI: 10.1201/9781003023401-5

Presentation Scientific Meeting:

> https://apha.confex.com/recording/apha/143am/mp4/free
> /4db77adf5df9fff0d3caf5cafe28f496/paper335049_1.mp4

Photo by Ketut Subiyanto from Pexels (https://www.pexels.com/photo/interested-multiracial-family-watching-tv-on-sofa-together-with-dog-4545955/)

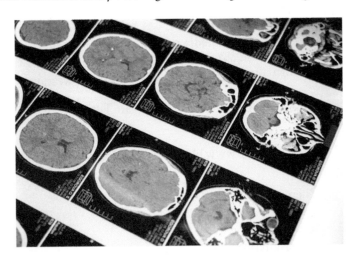

Photo by cottonbro from Pexels (https://www.pexels.com/photo/white-and-black-menu-board-5723883)

STORY SUMMARY

Before the advent of video streaming services, if you wanted to binge-watch a TV show you had to wait until a channel decided to run a marathon of episodes, often in the lead-up to the series finale. Netflix changed all that, and binge-watching is now something that people regularly do. But, as with all media consumption, there has been a growing interest in just what all that binge-watching does to us. That's the question at the heart of this *NBC News* story.

Reporter Danielle Page brings together a number of different sources to explain the different ways binge-watching might affect us – cognitively and emotionally. In the story she details how binge-watching can make us feel good and why that is, she then describes how the activity can actually help to alleviate stress, but also points out that there can be a bit of a "let down" once you burn through all the episodes. The story wraps up with some pointers on how to binge-watch responsibly.

WHAT IDEAS WILL YOU ENCOUNTER IN THIS CHAPTER?

- Multiple data sources may provide the foundation of many stories.

- Different data sources provide different types and amounts of information.

- Quotes from experts are not the same thing as a systematic study or review of evidence.

- Currency and novelty can drive news coverage of research.

WHAT IS CLAIMED? IS IT APPROPRIATE?

Brain chemistry changes, psychological responses and stress reduction can be associated with binge-watching television. However, the story also cautions the audience against binging to the exclusion of human interaction and recommendations about responsible binging are provided. Ultimately, more than one claim is discussed in this story.

WHO IS CLAIMING THIS?

This story focused on health impacts associated with media consumption, and the writer of the *NBC News* piece integrates disparate sources of evidence in support of viewing habits and outcomes.

WHY IS IT CLAIMED?

The consumption of television programs has been forever changed with the introduction of streaming services that allow you to view season's worth of programming in a matter of days. (Though, it should be noted, some streaming services are experimenting with uploading new episodes weekly, rather than all at once. Notable examples include Hulu's release schedule for The Handmaid's Tale and Disney+'s handling of its three Marvel series. Even with a trickling of the release of new episodes, you still have the option to wait until all episodes are released and then watch them in one sitting.) One of the affordances of a streaming service is that it can be done on a variety of platforms, ranging from smartphones to tablets to computers to the old viewing standard, television. Whereas, before, audiences were at the mercy of network and cable TV stations and could only watch a show on one medium – the television. The news story touches on the impact of a leisure behavior that is rapidly changing in society. While this story preceded the social distancing associated with pandemic concerns, it feels even more appropriate now than when the story first appeared in 2017.

Typical television consumption and binge-watching behavior provided the foundation for the story. The story then considered drivers of binge-watching along with explanations for the behavior. Claims included that binge-watching a series provides rewards that are reflected in brain chemistry and psychological responses. The story closed with a caution about binging to the exclusion of social interaction with real humans.

The story begins with a U.S. government survey that characterizes how Americans spend their time (spoiler alert: the average

American spends 2.7 h watching a day) and a corporate survey by Netflix that characterizes binge-watching behavior (another spoiler alert: 61% of survey respondents watch two to six episodes of a series when they sit down to watch). The story then explores possible explanations for binging, including physiological reasons (dopamine levels and lighting up pleasure centers of your brain) and psychological reasons (identification and ways we interact with shows and characters on shows). Stress reduction and connecting with others who obsess about the same series are also offered up as possible explanations. These explorations of the consequences of binge-watching were based on interviews with clinical psychologists and psychiatrists and a study of binge-watchers presented at an American Public Health Association conference.

IS THIS A GOOD MEASURE OF IMPACT?

Hours of television watching, a leisure time activity quantified in the American Time Use Survey (ATUS to its friends), seems like an appropriate endpoint of interaction with the medium. An interesting question is if/when the ATUS expanded its definition of television watching to include streaming series from vendors such as Netflix or Hulu. The binge-watching scale described below was a simple split of the time spent viewing a series each day. However, the impacts of these behaviors are not clearly evaluated.

HOW IS THE CLAIM SUPPORTED?

The claims regarding the impact of binge-watching TV shows are primarily supported by interviews with psychologists and a link to a presentation at a scientific meeting. Systematic studies of the impact of binge-watching series are not cited except for the results of a voluntary response survey that was presented at a professional conference. Assuming the experts quoted in the story are familiar and current with research related to the impacts of binge-watching, then their opinions might be viewed as informed and useful.

What Evidence Is Reported?

As noted previously, interviews with experts and a scientific presentation are the foundation of much of this story. Background about the patterns of television watching and binging behavior was determined by a government and an industry survey, respectively.

Reading Research behind stories: There is a hierarchy of information in stories – probability sample surveys and designed experiments are most likely to be representative and valid while convenience samples and expert opinions may be neither representative nor valid.

HOW MUCH TELEVISION DO YOU WATCH? GOVERNMENT SURVEY SAYS ...

What do Americans do on an average day? How much time do Americans spend at work, doing household chores or enjoying leisure activities like watching TV? The American Time Use Survey is a standard data source for understanding and characterizing the behavior of the free-range American aged 15+. This survey was started in 2003 and generates annual estimates from monthly samples that are collected throughout the year. Participants are selected as follows:

Start with the Current Population Survey (monthly survey of the U.S. labor force) household

AND THEN Select an ATUS household

AND THEN Randomly choose one individual aged 15 or older

AND THEN call this individual to get responses

The randomly selected individual is assigned a day (Monday–Sunday) to report their activity during a 24-h window the day before the call – here, a call on Friday would review the activity from 4 am Thursday through 4 am Friday, the day of the call.

The survey methodology section lays out all of these details including how often a household is called to get an interview (8 consecutive weeks) to how surveys were conducted

> Reading Research:
> Coding categories of responses is a difficult task and the best research establishes precise rules for such assignments.

(computer-assisted telephone interview) to how activity descriptions were coded (six-digit code from the "ATUS Coding Lexicon"). They then have to determine the rules for coding complex combinations of behaviors. For example, suppose you are listening to an audiobook while walking your dog. Is this reading or exercise or both or can one be considered a primary activity and the other a secondary activity? ATUS coders use a set of rules to figure this out.

After the data are collected and coded, the data have further processing that includes making informed guesses for missing data/imputing non-responses for items and weighting responses from each quarter to produce population estimates that are less biased. Even after all of that work, the sources of error are discussed. *Sampling error* reflects that different samples that are selected from the same population will differ; this is the error that you encounter in intro statistics classes. *Nonsampling error* reflects systematic differences between the sample estimate and the population value that might arise from nonsampling part of the population or nonresponse by some sampled individuals or data processing errors. The only result of this survey included in the story was the reported average amount of TV watched, which was 2.8 h per day. (It was also the leisure activity most enjoyed by respondents.) The careful reader may notice that 2.7 h per day was reported earlier in this chapter and 2.8 h per day was reported here. The 2.8 h per day average was based on the most recent ATUS results while 2.7 h per day was based on the ATUS results

> Reading Research:
> Sampling variability is captured in the margin of error estimates in survey research.

when the story was published. Questions you might consider if you were to explore this issue further include: What is the second most common leisure activity in the United States? What might similar time use surveys in other countries tell you about non-U.S. centric leisure activities?

ARE YOU A BINGE-WATCHER? INDUSTRY REPORT SAYS ...

Netflix is an online service provider that primarily streams video entertainment. But, not only does Netflix stream entertainment, it also collects data on what entertainment is being consumed and how people consume it. From the *Methodology* sec-

Reading Research:
It is fair to ask why a study was conducted. This was produced by the Netflix Media Center and while it summarized some viewing behaviors, almost half of the report was devoted to listing series that might be binged.

tion of the Netflix survey discussed in the *NBC News* story, we learn data from viewers residing in 190 countries during a window of time between October 2015 and May 2016 were analyzed. The flow of how a Netflix subscriber became part of the pool of people whose viewing habits helped define binging was:

Take all Netflix programs AND All Netflix consumers/ members

AND THEN Select 100 Series that are potentially binge-able

AND THEN Select consumers who watched ALL of the first season of a series

Data collected and analyzed for these individuals include days and hours to complete the series. It wasn't clear if a member completed more than one series if they would be included more than once in the data. The number of individuals who contributed data to this analysis also wasn't reported. It is worth noting that someone who binge-watches on another streaming service or digital video recorders (DVRs) all episodes of a series for later binge-viewing would not be counted here. Does the population of Netflix viewers look similar to these other groups? Do these Netflixers look similar to the population characterized by the ATUS?

The analysis reported that 50% of viewers who completed watching a series spent 2 h and 10 min or less per session and, further, 50% of viewers completed the series in five days or less. Interesting that this was per session. Can you imagine a problem if this was defined as "per day"? If someone started watching at 11 pm and finished watching at 1 am the next day, then this would contribute 2 h to a session of watching but 1 h to two different days. The report then defined "The Binge Scale" as

Savored: view < 2 h/day

Devoured: view > 2 h/day

So, what if you watch exactly 2 h/day unclassified? Probably not likely to see someone viewing exactly 2 h/day since viewing habits were likely recorded to the millisecond; it is reasonable to assume that a computer was measuring this viewing time. This report did not claim to be a scientific paper and you shouldn't expect it to be. While some background for the study was provided, there was a lot of information that would have been interesting to see. For example, how many people were studied? Which series was consumed the fastest? What are the differences between countries? (You might expect that much of this is analyzed by Netflix. Would this analysis be viewed as proprietary in Netflix and thought to be important for business?) What was the distribution of time to complete viewing each series? (Don't ever be happy with only a reported measure of a central value such as a mean or median from an analysis. A good summary will always include a measure of variability or distribution of responses along with a summary of a central value.)

IS WATCHING LOTS OF TV IS GOOD, BAD OR BOTH FOR YOU? EXPERTS SAY ...

Scientific research is published in journals and presented at professional conferences. This research often involves the systematic investigation into hypotheses about the pos-

> Critically reading stories: Experts can provide great quotes but are they an expert in the area discussed and do they have evidence in support of their assertions?

sible relationship between some exposure or behavior (e.g., binge-watching a TV series) and some response (e.g., stress, depression). Experts who are trained to treat people for particular kinds of behaviors – say, stress, depression or addiction – are not necessarily researchers. The psychologist and psychiatrists quoted in this story may be outstanding practitioners and therapists; however, you can't know whether they are speaking from a review of the literature in the field or if they are speaking from their specific years of experience treating a broad spectrum of patients. These experts provide potential explanations for why people binge-watch, they describe possible benefits and they discuss the potential costs associated with binge-watching behavior. It's all useful information, but knowing whether these reflections are based on anecdote or on research would be helpful as we decide how much we want to trust their insights.

An explanation or quote doesn't equate to a summary of the evidence in support of the explanation. It might. We can hope that the journalist confirmed this as part of background work for the story but we can't tell from the story.

BINGING AND STRESS? SCIENTIFIC PRESENTATION SAYS ...

A research study was cited in the news story to describe what happens after you finish binge-watching a series. Researchers from the University of Toledo presented at the 2015 meetings of the American Public Health Association. They described a study that involved more than 400 survey respondents.

Participant data was collected using Amazon's Mechanical Turk, or MTurk (https://www.mturk.com/). This is an online system where people can volunteer to participate in a research study for some small amount of compensation. The University of Toledo researchers used MTurk to take a cross-sectional look at what people report in terms of viewing habits and stress as well as other characteristics that might impact viewing behavior such as age and economic status. Three issues that immediately come to mind:

1. Respondents represent a sample but from what population? The population of people interested in online research using Mechanical Turk? These are volunteers and may or may not represent a population of interest or relevance to you.

2. Respondents self-reported their binging behavior. Does this mean that people might have very different definitions of a show-viewing binge? Absolutely.

   Reading Research:
   A standard and uniform definition of response categories will increase confidence that respondents are answering the same question.

3. The study was cross-sectional. Cross-sectional means that all variables including responses (e.g., stress) and potential predictors (e.g., binging) were measured at the same time. Can you say if binge-watching results in higher levels of stress or if higher levels of stress result in binge-watching? A formal study would be needed to disentangle the direction of this relationship or to understand if both stress and binge-watching were caused by some other factor. For example, would someone who felt socially isolated be more likely to binge and to feel stress?

The researchers acknowledge these limitations and emphasize that this study was exploratory and the results needed to be confirmed with future studies. The story did not capture any of this nuance or the limitations identified by the researcher.

What Is the Quality/Strength of the Evidence?

There really wasn't any evidence reported to support the claims of the psychological and physiological benefits of binge-watching or the downsides of binge-watching. Quotes from experts without some link to an evidence base are not convincing.

IS THE CLAIM REASONABLE IN ITSELF? DOES PRIOR BELIEF IMPACT MY BELIEF? CONFIRMATION BIAS?

Now you have watched the complete first season of a show or two, and have been classified as someone who "savored" a series. We bet that your prior belief was that becoming a series-binging zombie is probably

Learning more: Searches in web browsers can be a good way to find other projects related to some research problem. For example, a search for "binge-watching addiction" on scholar.google.com yielded 14,900 hits.

not healthy. Expert quotes generally supported this belief; however, as noted above, a quote from an expert or two should not sway you to take one position or another. This does not mean that binging is without either benefits or costs, simply that there is not enough here to get a good sense one way or

Reading Research: Consider where the research was published. Journals with connection to professional scientific societies likely merit more attention.

another. A quick search of the scholarly literature would be useful if you felt driven to see what systematic studies of the impact of binge-watching have been conducted and what was learned from them.

HOW DOES THIS CLAIM FIT WITH WHAT IS ALREADY KNOWN?

This question was not answered by the story nor by the background materials cited in the story. A quick search of health effects associated with television-viewing habits provides links such as obesity being a possible outcome, but other outcomes are possible. Investigating the background of a study can easily morph into a research question which spurs a whole new study.

HOW MUCH DOES THIS MATTER TO ME?

This story led us to reflect on our own viewing habits and to compare our viewing habits with those reported in the ATUS and Netflix reports along with personal reactions to consuming a series. We don't have a strong reaction to the claims that were based on expert interviews as our fear is that it may be possible to find an expert to express a particular opinion that runs counter to the scientific consensus. We are not saying this happened here, but we cannot be convinced by the assertions of the experts quoted in this *NBC News* story.

There are a lot of questions that surfaced and remain after reading this. What other factors drive binge-watching? Could an experiment be designed to randomize binge-watching to a group of viewers and non-binge-watching to another group and then observe the difference in stress responses or other psychological and physiological responses?

CONSIDERING THE COVERAGE

We've talked a bit about some of the holes we could poke in the *NBC News* article about binge-watching, but what we haven't discussed is what would make a news outlet want to cover this particular story anyway? In another chapter, we discuss the concept of news values – those things that propel news media to cover a story. Among them are timeliness, impact and proximity. Two at play in this story are less discussed, but often quite powerful – currency and novelty.

Uses and gratifications is a communication theory that assumes audiences are not passive consumers of media, but rather active users of media. It makes room for both agency and choice when it comes to media use and its effects, which theories of direct effects did not always do.

You have likely heard the phrase "Netflix and chill," though it's perhaps becoming dated at this point. But, it did have its moment as it seemed as though everyone was obsessed with some show on Netflix (or any of the other streaming sites). The ubiquity of that phrase seemed to signal that the concept of binge-watching had some currency – in a journalistic sense, the currency is the idea that something you report on is at the top of the mind of the public. "Netflix and chill" would suggest binge-watching was a public concern. Novelty enters the scene as the ability to binge-watch a series, as mentioned, is a relatively new thing. The currency and novelty of this story may translate into clickbait that induces casual readers to follow up on this headline and the quick summary of the story.

There is a long concern in media and communication studies about the way that media consumption affects us – early theories imagined that media had direct effects on users. That a producer could create a message, publish or broadcast it, and we would just take it in without really thinking too much about it. Other theories, such as uses and gratifications, attempted to make space for audience choice in our understanding of medium consumption. The *NBC News* article seems to be framing its discussion of binge-watching within a direct-effects understanding of media – an understanding that many media scholars have backed away from, but not necessarily the experts quoted in the article.

One expert, Dr Renee Carr, is quoted as saying "The neuronal pathways that cause heroin and sex addictions are the same as an addiction to binge watching ... Your body does not discriminate against pleasure. It can become addicted to any activity or substance that consistently produces dopamine." While there is research on the way the body physically responds to media consumption, the expert quoted here does not mention that line of work to support her claims. The reporter, in turn, does not seem to push the expert hard enough on how she can make such claims. The role of a reporter is to push a source because a journalist wants to make sure the information they share with their audiences is well-vetted and sound. Instead, the reporter just publishes what the expert says with no qualifications. In this instance, it would seem that the novelty of an expert saying binge-watching can be like using heroin overrides the reportorial impulse to push the expert to explain just how she could know that.

That's not to say that every time the news value of novelty, or currency for that matter, propels reporting that it results in a less than critical approach to the topic. But a careful reader, just like a reporter and a statistician, should approach every incredible claim with some healthy skepticism. And, quite frankly, the reporter should have reached out to media studies scholars to weigh in on their story. Media scholars with a wide array of interests have been exploring whether the type of direct effects suggested by some of the experts in this article is actually possible. What they've found is a bit of a mixed bag, with perhaps the most important thing to know is that most scholars have found that there are generally moderating variables that influence how media affect us. A story which quoted media scholars would have been a bit more even-handed and not as alarmist as this *NBC News* article at times is. Many of us are concerned with the way our immersion in media affects our mental health and well-being, so it behooves reporters to approach stories about the subject with care and nuance.

REVIEW

Binge-watching TV series may have benefits both physiologically and psychologically, though the caution about binging might be warranted. This story cited viewing habits based on a government survey, binge-watching patterns based on an industry study and impacts of binging based on quotes from behavior therapists and a professional presentation. However, the reporter did not engage with media studies experts who might have helped qualify some of the claims about the direct effects of binge-watching. Binge-watching is a novel experience and many of us find ourselves engaged in at times, but journalists, as well as researchers, have to be careful to not allow the novelty of the experience to wash away the nuance with which they should approach the subject.

STATS + STORIES PODCASTS

Scholars have been trying to suss out the relationship between media consumption and emotions for decades. Stats + Stories has featured interviews with a number of scholars exploring this relationship.

Comedian and BBC presenter Timandra Harkness talked with Stats + Stories about using humor to communicate statistical information to broad audiences (https://statsandstories.net/society1/rss2).

Jessica Gall Myrick got a lot of attention for her work on cat videos. She talked with Stats + Stories about why watching cat videos might be good for us (https://statsandstories.net/media1/2018/9/27/556dsf9h3y0a67padsfdt1zok9yf9a).

If you're interested in understanding the science of love, Stats + Stories' conversation with Ty Tashiro about relationship science might be right up your alley (https://statsandstories.net/lifestyle1/2018/8/2/reading-the-book-of-love-what-can-you-learn-from-relationship-science-stats-stories-episode-50).

For a dose of television history, producer Rick Ludwin talked about his work on network television (https://statsandstories.net/entertainment1/2018/8/2/what-do-seinfeld-the-tonight-show-and-statsstories-have-in-common-stats-stories-episode-7).

CHAPTER 6

Tracking the Spread of "False News"

Fake News: Lies spread faster on social media than truth does – March 9, 2018 (by Maggie Fox, NBC News). Story Source: https://www.nbcnews.com/health/health-news/fake-news-lies-spread-faster-social-media-truth-does-n854896

Journal Article:

Vosoughi S., Roy D., Aral S. (2018) The spread of true and false news online. *Science* **359**: 1146-1151. https://science.sciencemag.org/content/359/6380/1146.full

Supplemental Materials associated with the article: https://science.sciencemag.org/content/suppl/2018/03/07/359.6380.1146.DC1

DOI: 10.1201/9781003023401-6

Photo by Samson Katt from Pexels (https://www.pexels.com/search/social%20media%20addiction/)

STORY SUMMARY

Information spreads quickly on social media which has raised concerns about the veracity of that information. How can you know if something that goes viral is true? This story from *NBC News* chronicles the work of researchers tracking the spread of "fake news" on social media. The researchers published their findings in the journal *Science*, from which reporter Maggie Fox quotes with some frequency. Fox reports that the study found that lies spread more quickly and broadly than any other type of information on the social media site, Twitter. While there has been some desire to want to blame bots for the spread of disinformation online, Fox tells readers that the study shows that people are actually to blame for the spread of fake news. She also discussed the researchers' struggle over what terms to use – should they use the terms "fake" or "false" to describe the information they were tracking, for instance – in addition to the method the scientists adopted in their study.

Fox wraps up the story by suggesting that, even if bots aren't the main cause of the spread of fake news, their role should not be discounted. And, while fact-checking seems like one way to fight fake news, the story cautions restraint as the spread of fake news seems tied more to feelings and beliefs than truth.

WHAT IDEAS WILL YOU ENCOUNTER IN THIS CHAPTER?

- Data can be extracted from archived digital sources, including networks.

- Data analyzed could be features extracted and derived from a network of connections.

- Omitting some data and redoing the analysis can help explore the sensitivity of the results to other potential factors.

- Repeating a study with a different experimental design can help confirm or refute a result.

- The importance of interviewing to the journalistic verification process.

WHAT IS CLAIMED? AND IS IT APPROPRIATE?

False news will be retweeted more often than a true news story. In addition, false news will reach more people faster than true news. Finally, "cascades" of chains of tweets are longer and more common with false news. While the journal article places responsibility for the cascades more in human hands than automatic bots, the story motivating this chapter includes a caution from a computer scientist about the activity of bots on Facebook and Twitter. The author of an editorial that accompanied the scientific work that inspired the story is also quoted; they shed light on other factors involved in decisions to retweet, such as a confirmation bias where a reader accepts and retweets news consistent with their prior belief. Finally, the reporter echoes the suggestion of the scientists that more research is needed to understand false news. The reporter closes with a historical connection of legitimate newspapers arising "with ethics promoting objectivity and credibility out of the ashes of a boisterous yellow press."[1]

WHO IS CLAIMING THIS?

MIT Media Lab researchers provide the foundation of, and source for, these claims. In the abstract of their *Science* paper, they describe a study of "differential diffusion" of "verified true and false news stories." The reporter of the story based on this research provides background and inspiration for this research in the doctoral research of S. Vosoughi who was reacting to the false reports that surfaced after the bombings during the 2013 Boston Marathon. Vosoughi's epiphany and inspiration was the impactful presence of rumors on social media.

WHY IS IT CLAIMED?

Researchers conducted a study using 12 years of Tweets and more than 125,000 stories on Twitter. These stories were harvested from the Twitter verse and assigned to "true news," "false news" and "mixed" categories. This true/false/mixed assignment was based on evaluating each story using six fact-checking websites. The spread of news was evaluated by looking at the characteristics of the retweets of each news item.

One part of this research and the subsequent story was the decision to avoid the phrase "fake news" in favor of the category label "false news." The researchers did this to avoid the problem they noted of the label "fake news" being used by some individuals, particularly politicians, to describe news or assertions that differ from positions or claims they are making. If "fake news" had been used as the category description here, would some readers assume that the researchers had a particular agenda they wanted to promote?

Reading Research:
Consider if the meaning of particular concepts or labels has changed.

IS THIS A GOOD MEASURE OF IMPACT?

How would you characterize the spread of a tweet on social media? Would you use the number of retweets? How about a sequence of users retweeting an original news source? How about the total number of people who saw the Tweet? All of these are good ideas. The researchers formally defined several characteristics of the tweeting history of a news story. (If you could take a time machine back 15 years, typing the words "tweeting history of a news story" in any future piece of writing never would have been predicted!)

Consider how you interact on social media platforms? Do you follow individuals? If you use Twitter, do you retweet posts? What leads you to retweet a post? How would you describe how a news story moves through social media?

The foundational concepts for this analysis were (source phrases in quotes):

- News: "any asserted claim on Twitter."

- Rumor: "social phenomena of a news story or claim spreading or defusing through the Twitter networks."

- Cascade: "unbroken retweet chain with a common, singular origin."

The news was also classified into categories, including politics, urban legends, business, terrorism and war, science and technology, entertainment and natural disasters.

The properties of a cascade were the focus

Reading Research:
Definitions are critically important. Your colloquial sense of terms or concepts (e.g., "news") may be formalized and used differently by a researcher.

Reading Research:
Data may be defined in terms of networking of interrelated connections. This is a common feature of social media data.

of analyses in this research, and in fact, the comparison of the distribution of cascade properties between false news and true news was the key focus of this work. Cascade properties included:

- Depth: Number of retweets in a cascade (depth = 0 if no retweets – common value for John's Tweets.)

- Size: Unique users in a cascade – accumulates over depth layers.

- Breadth: "Maximum number of users involved in cascade at any depth."

Table 6.1 provides an example of a cascade started by User 0 – the original news source. Two users, User 1 and User 2, retweeted this news. The tweet by User 1 was retweeted by User 1.1, and User 1.1's tweet was retweeted by User 1.1.1. The path started by User 1 ended with User 1.1.1's retweet. User 2's post was retweeted by three users with retweets associated with one of these users.

At the start of the cascade, User 0 tweets a news story and the cascade has only one user (size = 1) and depth = 0. Next, Users 1 and 2 retweet this news. At this point, three users (size = 3) are involved in this cascade that now has a depth of one. Each retweet of previous retweets adds a level to the depth. The breadth, 2 at this level, refers to the number of users involved at a particular level of activity in a cascade. Follow the other rows in this cascade

TABLE 6.1   Example Cascade with Properties Reported (Inspired from Figure S7 in the Science article's supplementary material)

	Cascade				Depth	Size	Breadth
Original Tweet	News!/(Re)Tweet me!!! (User 0)				0	1	1
Retweet	User 1		User 2		1	3	2
Retweet	User 1.1	User 2.1	User 2.2	User 2.3	2	7	4
Retweet	User 1.1.1		User 2.2.1		3	9	2
Retweet			User 2.2.1.1		4	10	1
No more!							

to see what happened with this piece of news. To reinforce the relationships among users in this cascade, take a minute and sketch out the connection between the users in this example cascade. It should look like a tree with branches possible at each depth level.

This certainly doesn't look like the datasets you routinely see in an introductory statistics class. Here, the relationship among users is the data of interest and not the characteristics of a particular user. Thus, data here represents connections, a network or tree of connected tweets by users. In addition, users might be represented many times in the dataset, both as tweeters and as retweeters.

HOW IS THE CLAIM SUPPORTED?

Let's start with the reported evidence and then focus on the quality and strength of the evidence.

What Evidence Is Reported?

The researchers extracted more than 125 million English-language Twitter cascades resulting from ~2500 rumors between 2006 and 2017. These cascades involved about 3 million people, and the researchers removed retweets from computer programs/bots from the dataset. The origin of the cascade was classified as false/true/mixed news based on an agreement of six fact-checking websites: snopes.com, politifact.com, factcheck.org, truthorfiction.org, hoax-slayer.com and urbanlegends.about.com. The characteristics of false/true/mixed news cascades were then compared.

The original six-page science article provides the highlights of the work while the document of 50-plus-page supplementary materials provides background and detail, such as providing concept definitions and discussing the veracity of news sources. By selecting from the collection of all tweets in a window of time, the researchers could avoid the debate that might arise had a particular news source been selected. Figure S1 in the *Science* article's supplementary material summarizes a 2014 survey evaluating the difference in trust levels for news sources in different ideological groups. Trust in different news sources is tabulated in the article's supplemental section S2, and the veracity of statements is plotted versus trust in the news source. Here, a 2014 survey by the American Trends Panel asked a sample of American people with web access about how much they trusted various sources of news about politics and government. These sources included major news organizations, broadcast organizations and other sources. Participants answered a question about the overall trust they had in the source and evaluated them on a scale from consistently liberal to consistently conservative. Here, CNN was the source with the highest overall trust – 54% of respondents viewed this as trustworthy. Before you look at the supplement, why don't you make a prediction about some of the news sources that you follow? The noteworthy lack of consensus in these comparisons supported the need to use another strategy to establish the "truth" of news. The use of different fact-checking software was justified from this perspective.

Thus, each of the ~2500 rumors in the dataset launched one or more cascades of tweets and retweets. Up to 10,000 cascades could be associated with a rumor. Each cascade had its own features as well, including size and depth as described above. The cascades from Twitter required some processing to "infer the correct retweet path of a tweet" using information about follower relationships between users in a cascade along with time when retweets occurred. The researchers compared the distribution of different characteristics of the diffusion of news, such as the cascade features noted above, for false, true or mixed news. The scientists summarized these impacts, noting that "falsehood diffused significantly farther, faster, deeper and more broadly than the truth in all categories of information."

What Is the Quality/Strength of the Evidence?

Formal statistical hypothesis tests were employed to compare the measures of true and false news cascades. At a basic level, this is simply a two-group test for distributions of depth, size, virality and breadth between true–false news types. A similar strategy was used to compare the same traits between political and non-political news. The scientists also compared the time required to achieve certain impacts for tweets, such as the time to achieve a certain specified number of users.

The scientists also used a modeling technique, logistic regression, to examine the odds of retweeting as a function of users' characteristics (followers, followees = # of people you follow, age of Twitter account, engagement and verification of account by Twitter) and news veracity. One of the predictor variables, engagement, was defined as the ratio of account activity (tweets, retweets, replies, favorites) divided by Twitter account age (days). The analysis suggested that the odds of retweeting false news were 70% higher than the odds of retweeting true news when all user characteristics are held constant. It would have been interesting to see if the effect of news veracity was the same for user characteristics. For example, it may be that a new Twitter account is more likely to retweet false news than an older account (or the opposite). Expanding the model to investigate whether there's an interaction between news veracity and account age would allow this question to be addressed.

The analysis included a consideration of the emotional content of the tweets. A sentiment comes in different flavors of intensity. For example, "I believe you are mistaken" and "You are a clueless, unthinking idiot" both suggest disagreement, but there is a clear distinction between the vehemence with which it is expressed. Sentiment analysis would attempt to map degrees of direction and intensity to the language used. This is typically done after the so-called stop words (e.g., the, this, a, …) are removed from a sentence. These scientists used a list of 140,000 English words and their connection to eight emotions in their analysis of emotional content.

The novelty of news was also explored. This was examined by comparing a particular user's retweet to a sample of background tweets from people this user followed. Novelty was basically defined in terms of how far a tweet was from these sampled tweets.

IS THE CLAIM REASONABLE IN ITSELF? DOES PRIOR BELIEF IMPACT MY BELIEF? CONFIRMATION BIAS?

Our expectation, and we suspect we share this with the researchers and the reporter, was that the diffusion of false news would be different from that of true news. The results of this analysis confirm this belief. One aspect of the work of the scientists that lends support and credence to this work is the pains they took to evaluate how sensitive their analyses were to assumptions that they made. These analyses included:

- Recognizing that tweets were clustered for certain calculations.

- Conducting an additional analysis using human fact-checking to validate previous results.

- Comparing analyses with and without bots (and comparing bot detection algorithm).

The scientists work hard to extract reasons for the patterns they observe and the comparison of originators of news, sentiment, novelty and factors impacting retweeting demonstrates the desire to unpack the reasons for what is being observed. Finally, the scientists suggest the value of additional research such as behavioral intervention to change the spread of false news. (It may feel like every scientist says that more research is needed. Understanding and insight evolve over study with conclusions from one study often being explored in subsequent investigations.)

HOW DOES THIS CLAIM FIT WITH WHAT IS ALREADY KNOWN?

The scientists summarize previous research, noting that much of this early work has addressed the spread of a single rumor or multiple rumors associated with a particular event. Other cited research addresses detecting rumors, slowing the spread of rumors or a theoretical model for the spread of a rumor. The novelty of this project is that it "comprehensively evaluate[s] differences in the spread of truth and falsity across topics" and "examine[s] why false news may spread differently than the truth." Thus, this work dramatically expands the description of how true and false news cascades in social media and delves into the reasons for the differences that are observed.

As an aside, a replication of this study would be relevant. Would you see the same pattern of false-true news cascades in non-English tweets? Would patterns of "likes" in other social media patterns reinforce what was observed in this work?

HOW MUCH DOES THIS MATTER TO ME?

This story should cause a reader to think about what they might forward. Twitter is a space where you can communicate one to many. Unlike traditional broadcasting, there is no gatekeeper deciding what content is worth sharing with an audience and what content is not. (Traditional gatekeepers did this by weighing information's newsworthiness as well as by checking to make sure the facts held up under scrutiny. There are examples when accounts on Twitter are suspended but this is relatively rare.) If you have people following what you say, then you need to be careful what you retweet. There is a temptation to retweet a post that is consistent with your beliefs, but the story and underlying analysis may cause you to consider and check the veracity of a claim or story before retweeting.

CONSIDERING THE COVERAGE

One of the challenges that researchers often face is getting news media to cover their work, this is partly due to the fact that it can sometimes take specialist knowledge to understand a study and sometimes because it can be hard for a journalist to see what's newsworthy in a study. That is obviously not a problem in this case

– journalists have long been concerned with social media's ability to facilitate the spread of unverified information and, frankly, anything that suggests the importance of traditional media gatekeepers – such as journalists – is going to get a reporter's attention. What's more interesting, though, is how the reporter chose to cover the story. Rarely in a news article does a reporter extensively quote a research paper. The reporter might pull one or two quotes from the conclusion, but this story from *NBC News* quotes extensively from the journal article itself. That, in many ways, is a scholar's dream come true, but it's not necessarily a shining example of best practices in journalism.

As we've already noted, verification is the essence of journalism. This means a reporter's job is to not only report what someone, or some study, says but also look to other sources to verify the facts or to explore alternative interpretations of an issue. The reporter certainly seems to have gone to

> Interviewing is a crucial part of the journalistic process – it allows a reporter to verify the facts they have found in their reporting and to also push sources to provide evidence supporting particular interpretations of an event or an issue. A story that does not include interviews is underreported at best.

other sources to help make sense of the study's findings related to false news; however, it does not appear the reporter actually interviewed anyone in the course of their reporting. Instead, the journalist appears to have referenced past *NBC News* reporting on similar issues and to have quoted statements put out by other new media scholars about the findings of this particular *Science* publication. Maybe reading this you're asking yourself, "What's the big deal?" To which we'd respond that, in quoting statements and not conducting interviews, journalists are still acting more as stenographers than reporters. A reporter's job is to push a source to provide support for what they say. For instance, the reporter quotes one of the study's authors as saying in a statement that

"Twitter became our main source of news" in the aftermath of the Boston Marathon Bombing. Who is the "our" in that statement and how could the researcher know that's actually true? Research by the Pew Research Center (Geiger, 2019) actually shows that television remains the main way that Americans get news and, while internet sites are gaining ground, social media only account for 15% of Americans' news consumption. So, no, Twitter has not become "our" main source of news. Had the researcher interviewed Soroush Vosoughi, the quoted author, they might have been able to push back on that assertion and got the researcher to say something either (a) they could know or (b) explain where the information to support the assertion that Twitter is a major news source comes from.

Towards the end of the story, the reporter seems to pull in an outside perspective on the study, quoting researcher Filippo Menczer about how bots *do* fuel the spread of misinformation in social media. However, this, too, came from a statement rather than an interview. But we do not know if this was an email statement sent to the reporter or a statement Menczer issued publicly about the findings of the *Science* study. The other "outside" expert the reporter quoted in the story actually came from a policy paper published in the same issue of *Science* as the study the journalist is reporting on. It would appear that the reporter conducted no actual interviews during the reporting of this story, relying on the research article and statements and papers published by other scholars to help them make sense of the study's findings. At no point was the reporter ever in a position to push any of the experts to state how they could know what they claim to know. That's just not good reporting. The original source article in *Science* was targeted at the members of the American Association for the Advancement of Science. Note that our observations are not critical of the original source but of how that source is being explored and communicated to a general audience.

The bones of an interesting news story are all here. The reporter does a good job of contextualizing the study's findings, but the verification is missing. It is difficult to verify the facts when you don't talk to sources. Unfortunately, a lot of the news coverage of this study followed a similar pattern – quoting extensively from the study or from official statements, but not seeming to interview experts during the course of the reporting. One exception is a story published by Vice News. In the story (Marrelli & Zimmer, 2018), the reporter interviewed not only study coauthor Soroush Vosoughi but also an outside expert on the issue. The interview with Vosoughi gave a bit more background on what propelled the study on fake news on Twitter, while the outside expert – Joan Donovan of the Data & Society research group – helped the audience better understand whether the findings of Vosoughi's study were anomalous. Donovan is quoted as saying,

> [The study] confirms much of what we are seeing in social media overall, which is that headlines sell the story and false news is both less expensive to make and more interesting to read, as has always been the case with tabloids.
>
> (MARRELLI & ZIMMER, 2018)

Importantly, the outside expert also contextualizes the study outside the realm of new media, reminding the reader that sensationalistic, fake news stories have been the trade of some media for a very long time. Donovan also alerts the reader to the possibility that there might be a limited shelf-life for the rapid spread of fake news, noting that in her own research she's found that window to be 12–24 h. Though other experts quoted in the Vice article do point out that the resharing, or retweeting, of fake news or other types of misinformation can lend its credibility that outlasts that short shelf-life.

The difference between the two stories is striking – the NBC story feels like a play-by-play of what the study found while the Vice article attempts to unpack what the study's findings actually mean. If journalists are true to serve the communities they cover, then they must do more than simply recite what a study, or a source, says. A story about fake news is fairly low stakes, but studies about things like vaccinations are not, as we have all witnessed in the "controversy" over the (non) relationship between vaccines and autism and more recently in some of the coverage of the COVID-19 vaccines. Responsible reporting, particularly when it comes to scientific studies, must include interviews if it is truly to serve the public good. Otherwise, a news outlet might as well publish the press releases they get from journals or researchers and call it a day.

REVIEW

News is framed as a social media claim in this headline, story and background research. A news item is evaluated for veracity, and its diffusion throughout a social media network is characterized and quantified. The distributions of diffusion network traits are compared between false and true news items, and models to examine factors leading to the retweeting of news are explored. A study based on analyzing a full archive of a social media platform is a compelling foundation for analysis and a news story inspired by scientific analysis. However, journalistic coverage of such a study should interview sources about the study's findings to get clarity on how things were conceptualized as well as rule out other interpretations of the findings.

STATS + STORIES PODCASTS

Stats + Stories has featured a number of conversations about false or fake news and its propagation.

The conversation with Filippo Menczer focused on the spread of fake news across networks (https://statsandstories.net/media1 /2018/8/2/if-your-friend-believes-it-then-it-must-be-true-track-ing-the-spread-of-fake-news-across-networks-stats-stories-epi-sode-28).

Steven Lloyd Wilson described social media impacts on democracy (https://statsandstories.net/politics1/social-media-data-and-democracy, and Joshua Tucker discussed an investigation into detecting the presence of Russian bots and their potential impact https://statsandstories.net/politics1/how-to-identify -russian-bots).

The detection of fake social media accounts was one of the topics discussed by Mark Hansen (https://statsandstories.net/ media1/2018/11/15/understanding-data-in-the-digital-age-stats-and-stories-episode-70).

NOTE

1. In the United States, the term "yellow press" or "yellow journalism" has been used to label reporting that was sensational in nature and which sometimes exaggerated or made up the facts in a story. Journalists engaged in such reporting often engaged in unethical practices and, when it came to crime reporting, sometimes got in the way of actual investigations.

CHAPTER 7

Modeling What It Means to "Flatten the Curve"

Why outbreaks like coronavirus spread exponentially, and how to "flatten the curve" – *Washington Post* – March 14, 2020 (by Harry Stevens). Story Source: https://www.washingtonpost.com/graphics/2020/world/corona-simulator/

Supplemental Source: Coronavirus: What is the R number and how is it calculated? https://www.bbc.com/news/health-52473523

DOI: 10.1201/9781003023401-7

Photo by Yaroslav Danylchenko from Pexels (https://www.pexels.com/photo/woman-in-brown-dress-holding-white-plastic-bottle-painting-4113084/)

STORY SUMMARY

In this story, the *Washington Post*'s Harry Stevens works to help the audience understand how COVID-19 spreads in a community and also explains what it means to flatten the curve. Unlike other stories about these issues, Stevens does not write a long text story about this and he is not echoing or interpreting research publications about the spread of a virus in a community; rather, he produces a series of interactive graphics to help tell the story of the virus's spread. Throughout the story, key terms the audience needs to understand are bolded with an interactive graphic following soon after the term's appearance, so that the article not only *tells* the audience what something is but also works to *show* them as well. Generally, humans process visual information more quickly than they process textual information, so it's an interesting bit of reporting that relies on the strengths of both the text and the visualizations to communicate the story to the audience.

WHAT IDEAS WILL YOU ENCOUNTER IN THIS CHAPTER?

- Simplified representations of a complex problem can capture the key features of the problem.

- Visualizations of a concept may be the best foundation for a story.

- The importance of simplicity in crafting a powerful message.

WHAT IS CLAIMED? IS IT APPROPRIATE?

A number of claims are embedded in this story. First, outbreaks of viruses grow exponentially implying that infections might spread completely and quickly through a population. Exponential growth implies that the increase in the number of infected is proportional to the number who are currently infected in the population. You encountered this proportionality number during the COVID-19 pandemic when there was a lot of discussion of the so-called R value. The BBC story dives into a description of R. This quantity captures how many people on average are newly infected by virus-carrying individuals. Here, when $R > 1$, the virus presence is growing in the population, when $R = 1$ then the population is staying constant and when $R < 1$, the virus is dying out.

Consider what happens when $R = 2$. Suppose we start with only a single individual with the virus. This person infects two and then these two individuals infect two and this continues. So the pattern of the number of newly infected individuals is 1-2-4-8-16-32-64-128 and so on. Now, this ignores the nuance that individuals are not contagious forever and that most will recover and not infect other; however, exponential growth with $R > 1$ is like a wildfire that will consume all of the fuel (uninfected) until no fuel remains.

The impact of forced quarantine can protect a separated population, but once the wall between an infected and quarantined population is breached, all bets are off and an unexposed population will experience the same pattern of exponential growth in infections. Finally, two illustrations of how degrees of social distancing impact the number of infections over time are included.

WHO IS CLAIMING THIS?

Harry Stevens, a graphics reporter at the *Washington Post*, wrote a story that featured computer simulations that serve as the foundation of the story. This is not based on some background scientific paper or government report; rather, the story reflects Stevens' journalistic understanding of the underlying science behind the spread of viruses and control strategies to mitigate their spread. This article provides an exposition of an important concept that will equip a reader to consider the impact and control of a real pandemic virus.

WHY IS IT CLAIMED?

During the global COVID-19 pandemic, public health officials were clear and strong advocates of social distancing. They provided an early, critical message of the need to "flatten the curve" to avoid overwhelming hospital resources for intensive care of the most impacted by the diseases resulting from this virus. Stevens' story was not a new claim, but a clever visualization of this pattern which provides a sense of impact. The writers and coders of the simulation, and its accompanying illustrations, wanted to communicate the essence of how an infectious disease spreads and the time course of the disease in terms of how many in a population are infected at any given time and how many in a population have recovered.

IS THIS A GOOD MEASURE OF IMPACT?

The impact here was defined by tracking the trajectory of members of a population who could be classified into three distinct groups (healthy, sick/infected and recovered). The relative sizes of these three groups were tracked under differing conditions explored in this story (freely moving, quarantined, social distancing with three of four people adhering to distancing guidelines, and social distancing with seven of eight people adhering).

Impact is measured by the number of people who are infected with *Simulitis* at any point in time. By calling this simulated disease *Simulitis*, the reporter removes the explicit connection to COVID-19 and allows the reader to reflect on critical concepts in the context of the impact of a pretend virus on a simplified, abstract population. Extracting the key features of a situation and exploring and manipulating these features are critical for gaining insights from a simulation. Here, this won't be a perfect rendering of how the infection of a virus moves through a population but the essence of how this proceeds. The assertion that models are caricatures of reality that emphasize certain traits over other traits is common and applies here.

The main features monitored in this simulation are the number of healthy, sick and recovered over time. This is directly relevant and of immediate interest since hospital resources are often measured in bed capacity and the number of extremely ill infected people are likely to need these resources.

HOW IS THE CLAIM SUPPORTED?

Simulations of free-range humans (no restrictions on movement), quarantined subpopulations and population members practicing distancing were implemented to explore patterns associated with the number of healthy, sick and recovered.

What Evidence Is Reported?

The driving force of this story is simulated data. In essence – it's a story about disease spread based on made-up data derived from a simple model for what happens in a population. Basically:

Healthy ->

Infected ->

Recovered

Essentially this is a computer experiment where a small population of 200 healthy people has a single case of the disease *Simulitis* introduced. The computer experiment or simulation study is world-building at its best – anyone ever played Simcity or Animal Crossing? Simulations generate data from a set of assumptions. Here, the assumptions included:

1. If you are healthy and you contact an infected person, then you will get infected.

2. After a certain period of time, you will recover.

3. If you recover, then you will not get the disease again.

4. If you are not staying stationary, you start moving in some random trajectory until you encounter another individual or the boundary of your town where you bounce off these encounters back into the town. This is similar to a pool/billiards table where balls once moving do not cease moving.

This simulated world here was a rectangle with balls moving around – picture in your mind a billiards table with no pockets. The balls are one color when healthy, another when infected and a third color when recovered. These balls move around this rectangle until they encounter another person or they hit a wall. This model could just as easily represent molecules moving in a chamber as people in a population. The power is in the abstraction and the insights it might produce about the system that is represented by the simulation.

What Is the Quality/Strength of the Evidence?

The simulation and its assumption don't have to be an exact representation of what we would expect in a population. In fact, we know that you may or may not get sick if you interact with an infected person and you probably don't move around your community like a physics-defying billiards ball. Further, we know the time until recovery differs between individuals (some die as well). Finally, we don't know if you can get the virus again after you have recovered from an infection. These exact details don't matter for the purposes of this illustration.

The key features of how a virus is spread in a population and how control measures might impact this spread are captured by this simulation. Two great quotes (from sources that we can't remember) about models are that models are "caricatures of reality" and that they are "purposeful representations of reality." In the same way that a cartoonist at the fair captures and often accentuates the key features of a person they are sketching, a model captures the essential features of the system, here a pandemic-impacted population, that the modeler wishes to explore. (Your homework is to find the quote by the famous statistician George Box about model correctness. If it takes you more than one browser search to find this quote, then you typed something wrong.)

> Reporting and models: Sometimes computer models are created to explore relationships. These may be **deterministic** in that if you provide the same input, then you get the same output. Alternatively, models might be **stochastic** and provide the same input, then you get potentially different output. What type of model was in this story?

The old adage that "a picture is worth a thousand words" is embraced as part of this story. In Figure 7.1a, the time-ordered trajectory of the relative numbers of individuals in one of three states (healthy, infected or recovered) in a hypothetical population of 200 is displayed.

Figure 7.1b provides complementary information showing the relative share of the population in the three conditions. The peak number of infected cases are observed at time = 5 (units really don't matter here) based on the height of the infection curve in Figure 7.1a or the largest size of the infected slice in Figure 7.1b. The original visualizations in the story are dynamic displays similar to Figure 7.1b.

Figure 7.2a and 7.2b displays the same information for a population exhibiting social distancing. Here we see that the peak number of infected is much lower than in the condition when no social distancing is followed.

The key observation is that the population with no social distancing peaked earlier (time = 5) and had more cases at this peak (more than 150 of the 200) when compared to the population with social distancing (peak time around 14 with 100 of the 200 infected). The story is that the lower peak with social distancing would prevent the overloading of hospital resources and the later

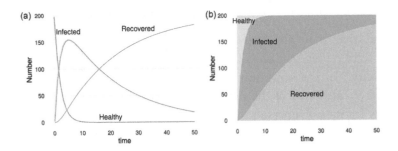

FIGURE 7.1 (a) Pattern of number of healthy, infected or recovered observed over time in a population not following any social distancing. (b) Relative number of healthy, infected or recovered observed over time in a population not following any social distancing.

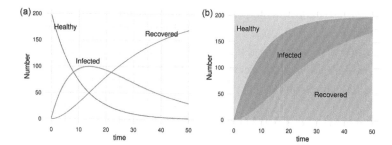

FIGURE 7.2 (a) Pattern of number of healthy, infected or recovered observed over time in a population following some social distancing. (b) Relative number of healthy, infected or recovered observed over time in a population following some social distancing.

peak suggested health officials would have more time to prepare to react to the pandemic.

IS THE CLAIM REASONABLE IN ITSELF? DOES PRIOR BELIEF IMPACT MY BELIEF? CONFIRMATION BIAS?

The story from public health officials was clear, you will "flatten the curve" – decrease the maximum number of cases at a particular point in time and delay the occurrence of the peak – if you enact certain control strategies such as social distancing. The challenge for us was picturing ourselves as a point on these curves and to think about what condition we could implement to impact these curves. This story translated the movements in a hypothetical population of infected individuals into realized values on these curves displaying infected counts. Thus, the impact of this story was to make real the public health advice, and the impact of following this advice, on a population. It is amazing to see that data can be constructed to represent the expected patterns in a real population.

HOW DOES THIS CLAIM FIT WITH WHAT IS ALREADY KNOWN?

Impacts of social distancing have been observed in the 2020 COVID-19 pandemic and the 1918 Spanish flu epidemic. Countries and cities that implemented stronger stay-at-home and social distancing rules had lower case counts and mortality in their communities. This story and its simulation-based illustrations extract from the real-world observed patterns of the trajectory of a virus and display results in a simpler computer-generated world.

HOW MUCH DOES THIS MATTER TO ME?

The displays of complex systems in a manner that is understandable and brings an "aha" moment to a reader are to be celebrated. This story and its embedded simulation do exactly that, and it is no surprise that this is one of the most read *Washington Post* stories ever published. What is your response to this simulation? Did the simulation help you better understand the impact of social distancing? Would you change your behavior given a better understanding?

CONSIDERING THE COVERAGE

There is a saying heard in newsrooms across the country when an editor is trying to get a reporter or an anchor to get to the point of something. "Please," these editors say, "please use the *fewest most powerful words possible* to tell your story." Journalists, particularly those generalists who are tasked with telling a specialized story, sometimes lean into over-writing as a way to mask inexperience with something. The reporter includes too many details, too much jargon, all to cover up for their own lack of expertise – often making what is already a complicated story that much more confusing.

While Harry Stevens' reporting on flattening the curve is a nice example of how visualizations can aid in storytelling, his writing is also impressive. He uses everyday language to explain complicated concepts – telling the story the way he might to his grandmother or to a friend over a beer in a bar. Being able to do that is a skill that journalists need to develop if their reporting is to ever have any real impact on the broad public. For example, when explaining the exponential curve in relation to infectious disease, Stevens writes this "has experts worried. If the number of cases were to continue to double every three days, there would be about a hundred million cases in the United States by May. That is math, not prophecy." While we like the humdinger about math and prophecy, the more important bit is how he explains what an exponential curve means for American lives. It also sets up the rest of the discussion later in the story about the various strategies that officials attempted to adopt early in the pandemic in order to curb COVID's spread.

One of the particularly difficult things for journalists to parse early on in the pandemic was what experts meant by the phrase "social distancing" and why they thought it was an important strategy to adopt in order to slow the spread of COVID. Before a series of visuals showing the impact of different social distancing strategies, Stevens writes, "health officials have encouraged people to avoid public gatherings, to stay home more often and to keep their distance from others. If people are less mobile and interact with each other less, the virus has fewer opportunities to spread." The second sentence there, about mobility and interaction, is modeled in Stevens' visualizations, but it is also important to spell it out for readers in black and white text. It's by avoiding interacting with others who might have COVID that we slow the spread of the disease – both visualizing that fact and stating it plainly helps reinforce that message for the reader.

Other news outlets were also working early in the epidemic to help their audiences understand what it means to "flatten the curve" – although not in quite the same way as this piece from Harry Stevens.

A story from the *New York Times* (Lai & Colllins, 2020) published a few days after Stevens' article also featured visualizations of curves – though they were not interactive. The visualizations were focused on understanding which countries had been most successful in flattening the curve to that point in time (the story suggested that Norway, Japan and South Korea seemed to be doing the best). But, the story was just a rehashing of the confirmed cases of the virus in the countries profiled, with no real explanation of what it means to flatten the curve or what it might demand from those reading the story.

KPBS, a public radio station out of San Diego, the day before published an article attempting to explain what both "flattening the curve" and "social distancing" meant (Chatlani, 2020). Social distancing, the article notes, "is exactly what it sounds like. It's keeping your distance – at least 6 feet – from other people." Just under that explanation was a simple graphic illustrating how the virus was spread by people sneezing in a space when close to others. As Stevens attempted to discuss in his article, the KPBS piece also tries to explain what it means that something is spread exponentially in an accessible language.

> many people may not realize they're spreading the virus. That's why officials say if there isn't social distancing, the number of people getting sick will grow exponentially like this:
>
> "If you take a single drop of water and you double the size of it every minute, within less than an hour it will fill a baseball stadium," Kaiser said.
>
> So, for every person who tests positive, another 2 people could get the virus, and those numbers keep doubling. This rate of this growth can quickly become a problem, because as the number of people getting sick goes up, so will the number of people who need to go to the hospital.
>
> (CHATLANI, 2020)

The KPBS article also included a curve flattening visualization, but as with the *New York Times* article, it was not interactive in nature and served as more of an illustration than anything else. The story ends with the reminder to "remember, flattening the curve isn't about panicking, just about keeping a safe distance to slow the rate of the virus." It is not as data rich or driven as

Stevens's article, but it succeeds where Stevens also did – in telling a complicated story in an accessible, easy-to-understand manner. Journalism is ultimately meaningless if the language is so complicated or convoluted that it turns an audience away. Thoughtfully using the fewest most powerful words possible to tell even the most complicated statistical stories can help ensure a story is read and, more importantly, understood.

REVIEW AND RECAP

An effective simulation, a virtual game, provided meaningful insight for connecting readers to concepts that were critical for understanding the impact of a pandemic virus in a population and suggested behavioral controls, here, social distancing. But, fancy visualizations don't override the need for communicating a story in a straightforward, easy to understand manner. There can be poetry in simplicity, being able to communicate something succinctly can help ensure both you and your audience understand what's at stake.

COVID CODA

We began work on this book before COVID emerged, began revisions as the pandemic seemed to reach a major apex and are finalizing our work on this volume as the world stares down the end of 2021 framed by the Omicron variant. As we contemplate another winter masked up and socially distanced, we find ourselves meditating on the simplicity of Harry Stevens's visual explanation of what it means to flatten the curve. Omicron is moving through communities with lightning speed, meaning we all have to be working to flatten the curve before it begins to even grow. More than two years into the pandemic, the work of reporters like Stevens continues to stand out for its ability to communicate complicated public health information simply. Even if not everyone follows the advice to wear a mask and maintain social distance, no one can argue that the information on why we need to do that isn't out there. During interviews with statisticians and public health officials for Stats + Stories, what we've been struck by time and again is that those experts generally feel the news media have done a pretty good job during the pandemic. There have been occasional missteps, as one might expect, but our Stats+Stories guests generally think reporters have fulfilled their duty to serve the public good when it comes to COVID. We think that's worth celebrating as we ponder what Omicron means for the pandemic and what might come next.

Bonus story:

> Headline: Disease modelers are wary of reopening the country. Here's how they arrive at their verdict. – *Washington Post* by Harry Stevens and John Muyskens May 14, 2020.
>
> Citation Story Source: https://www.washingtonpost.com/graphics/2020/health/disease-modeling-coronavirus-cases-reopening/.

STATS + STORIES PODCASTS

Stats + Stories has featured a lot of conversations about the visualization of statistical information, ranging in focus from its historic roots to current best practices.

Harry Stevens, the author of the headline, discussed this work and the use of simulation and visualization in data journalism (https://statsandstories.net/media1/the-most-viewed-washington-post-article-ever).

Amanda Makulec discussed the importance of data visualizations in understanding the story of COVID-19 early on in the pandemic (https://statsandstories.net/health1/coronavirus-visualizations?rq=visualization).

Early examples of data visualization are the topic of a conversation with Alison Hedley, who discusses the way visualizations have been used to understand population statistics (https://stat-sandstories.net/methods/19th-century-data-visualization?rq=visualization).

The importance of creating beautiful, easy-to-understand data visualizations in journalism was the focus of an episode featuring Alberto Cairo (https://statsandstories.net/media1/2018/8/2/visualization-and-reporting-goals-truth-function-beauty-insight-enlightenment-and-morality-stats-stories-episode-42?rq=visualization).

CHAPTER 8

One Governor, Two Outcomes and Three COVID Tests

Gov. Mike DeWine of Ohio Tests Positive, Then Negative, for Coronavirus. *The New York Times* – 6 August 2020 with 7 August 2020 update (by Sarah Mervosh). Story Source: https://www.nytimes.com/2020/08/06/us/mike-dewine-coronavirus.html?campaign_id=2&emc=edit_th_20200807&instance_id=21064&nl=todaysheadlines&regi_id=69604087&segment_id=35491&user_id=96153b730df88cfb6f14172c669a362e

Medical Journal Background Software: **Interpreting a COVID-19 test result** https://www.bmj.com/content/369/bmj.m1808

DOI: 10.1201/9781003023401-8

Photo by Jakayla Toney on Unsplash (https://unsplash.com/photos/zRGqHTHP-HQ)

STORY SUMMARY

In August of 2020, journalists reported that Ohio Governor Mike DeWine tested positive for COVID-19 in advance of a visit by President Donald Trump to the state. DeWine's positive result meant he could not meet with the president during the visit. Instead, he and his wife Fran holed up as they awaited the results of the second round of testing and as Lieutenant Governor Jon Husted took over the meet and greet duties. After that first initial positive, DeWine tested negative. And then he continued to test negative, leading to wild speculation about everything from the governor's feelings about the president to the trustworthiness of COVID tests to whether maybe COVID was a hoax after all.

In the *New York Times* story, reporter Sarah Mervosh described what happened and then worked to explain the intricacies of COVID testing, describing the different types of tests and how they might lead to conflicting results. This took place against the backdrop of Ohio purchasing some 4 million antigen tests – the kind that led to DeWine's false positive – in an attempt to expand testing in the state. Access to testing has been a major story of COVID in the United States, and the reporter took some time to also discuss that issue as well.

WHAT IDEAS WILL YOU ENCOUNTER IN THIS CHAPTER?

- Screening tests all have errors – sometimes people without disease will test positive (false positive errors) and sometimes people with the disease will test negative (false negative errors).

- The probability of a positive test result among people who have the disease is different from the probability that a person has the disease among people who have a positive test result.

- A positive test result in a population with a small number of people with the disease may still mean that it is much more likely for a person to not have the disease.

- The ability of conflict to drive a news narrative.

WHAT IS CLAIMED? IS IT APPROPRIATE?

Mike DeWine, Governor of the U.S. state of Ohio, tested positive for COVID-19 on a rapid antigen test that resulted in DeWine missing an opportunity to meet with the U.S. president who was visiting DeWine's state. After DeWine was tested again using a slower (days to obtain a result vs. minutes), more accurate (RT-PCR) test, he was negative for COVID-19. A replicate test administered a day later also was negative. The claim associated with this story reflected that screening tests are not perfect and different testing systems have different performance characteristics.

One focus of this story was the ability of testing to help with the control of a pandemic outbreak. As reported in the story,

> Experts are increasingly arguing that the best chance to catch the most outbreaks is through large numbers of less

accurate tests. But there are drawbacks: Antigen tests will miss some people who would test positive by P.C.R., with some past antigen tests missing up to half the infections they looked for.

WHO IS CLAIMING THIS?

The governor's office, likely through a public information officer in the office, released the news of the initial positive test result along with the subsequent negative test results. This story is newsworthy because (a) it involves a public figure, (b) DeWine had taken a somewhat aggressive approach to curb the spread of COVID compared to other states and (c) the positive test kept the governor of a battleground state from meeting with the American president running for re-election at the time.

WHY IS IT CLAIMED?

The data spoke for themselves in this story. Three tests administered in a short window of time produced two diametrically opposed results. There were two biological outcomes evaluated in the tests, one test detected protein segments and another probed for virus-specific genetic material. DeWine's positive, and then negative, status was determined by these tests. There has been a lot of local and national news coverage explaining the different types of COVID tests, with many trying to explain the different ways the tests work. Links to some of such stories are provided at the end of this chapter.

IS THIS A GOOD MEASURE OF IMPACT?

Seeing different results for repeated testing of an individual where little time has elapsed between testing was an ideal measure of impact, particularly the impact of errors in the results of screening tests. No test is perfectly accurate. Accuracy can be defined in different ways, particularly in ways that reflect errors in decisions that are made. Two simple errors are commonly used when describing screening tests – saying someone has a disease when, in truth, they don't (sorry Governor DeWine) or saying someone is disease-free when, in truth, they have the disease. Governor DeWine had three COVID-19 tests – the first rapid test was positive, and the second and third tests were negative. Thus, we assume his true health status is disease-free, particularly given the absence of other symptoms.

Screening errors can arise in a host of situations. Suppose a drug test is required by your employer. If you are drug-free but the test result is positive, then a false positive error may impact your employment. (An employer might require multiple positives before such a dramatic impact on employment status.) If you are a drug user but the test result is negative, then the employer may have a concern.

These types of screening errors are not limited to blood tests for disease or drug use. Facial recognition software not only unlocks your phone but might be used in cities or airports to identify criminals. If you are innocent of any crime but arrested or put on a no-fly list, then a false positive error has serious personal consequences. If you are a criminal, then a false negative may place the public in danger.

HOW IS THE CLAIM SUPPORTED?

False positive and false negative errors are the bad news in the screening test story. The correct decisions in screening tests – test result negative in a disease-free person and test result positive in a person with the disease – are the good news. The table below captures how the combination of true disease status (columns) combined with observed test results (rows) defines all of the possible outcomes for a screening study.

	TRUE Disease Status	
	Disease Present	**Disease-Free**
Test +	Correct!	False Pos. error
Test -	False Neg. error	Correct!

It is important to recognize that these errors can only be made when testing distinct groups of people. A false positive error can *only* be made when testing disease-free people. A false negative error can *only* be made when testing people with the disease. This differs from the real questions people (including Governor DeWine) want to ask.

"Do I have the disease if I test positive?"

"Am I disease free if my test is negative?"

What Evidence Is Reported?

Probability calculations can be used to calculate the probability of disease given a positive test result if you know the false positive error rate, the false negative error rate and the percentage of the population with the disease. The *British Medical Journal* (BMJ) provides a nice web calculator for looking at these probabilities along with a nice visualization of the disease and testing status of a hypothetical population. This calculator allows you to specify test false positive and false negative probabilities and then presents the probability that a randomly selected person from the population has the disease. In addition, this app shows the probabilities in terms of counts of a hypothetical population of 100 individuals classified into four groups based on true disease status (disease, no disease) and screening test result (positive, negative).

These probabilities may not be known and can be really hard to estimate – particularly when changing. In real life, there are serious challenges in estimating the probabilities/rates that get fed into calculators such as this – but that's beyond scope of this chapter. Some lingering questions might include: How many people had COVID-19 in Ohio when Governor DeWine was tested? How many did not have COVID-19? What were the rates of false positive and false negative errors for the tests?

What Is the Quality/Strength of the Evidence?

These error rates vary between different test types and even for tests of the same type. Richard Harris national public radio (NPR) reported that PCR false positives from the PCR test were approximately 2%, with variation attributable to the laboratory conducting the study and the test. National Public Radio reported that one rapid COVID-19 test had a false negative error rate of approximately 15% while better tests have false negative tests less than 3%.

IS THE CLAIM REASONABLE IN ITSELF? DOES PRIOR BELIEF IMPACT MY BELIEF? CONFIRMATION BIAS

The examples below illustrate a comparison of tests with different accuracies/error rates in communities with different percentages of the population who have COVID-19. This percentage, sometimes called disease prevalence, has varied over time in Ohio; however, it increased dramatically at the end of 2020. We illustrate the interpretation of positive test results in a comparison of two tests of differing accuracy in two populations – one with a lower rate of COVID-19 (say 2 in 100) and one population with a higher rate of COVID-19 (say 20 in 100).

COMMUNITY WITH LOW RATE OF INFECTION

Our local paper reported in August 2020 that 1.4% to 1.8% of donors to the American Red Cross had COVID-19. We round this to 2% for our calculation and consider a hypothetical population with 100 people, only 2 people in the hypothetical population would have disease and 98 would be disease-free.

Rapid, Less Accurate Test

Suppose we had a rapid test with a 10% false positive error rate, 15% false negative error rate to a population where 2% of the people are truly positive. With the error rates described for this test, both of the people with disease test positive and 10 of the

98 disease-free people test positive. Based on this, a person with a positive test (2 + 10 = 12) has a 16% (2/12) chance of having the disease.

	Disease Present	Disease-Free	Total	
Test +	2	~10 =98*0.10 (false pos. error rate)	12	2/12 = 16% chance of disease with + test
Test -	~0 =2*0.15 (false neg. error rate)	88	88	
	2	98	100	

Slower, More Accurate Test

Suppose we had a more accurate test with a 2% false positive error rate, 1% false negative error rate that is applied to this population where 2% of the people are truly positive. As noted above, only two people in the hypothetical population would have disease and 98 would be disease-free. With the error rates described for this test, both of the people with disease test positive and two of the 98 disease-free people test positive. Based on this, a person with a positive test (2 + 2 = 4) has a 50–50 (2/4) chance of having the disease.

	Disease Present	Disease-Free	Total	
Test +	2	~2 =98*0.02 (false pos. error rate)	4	2/4 = 50% chance of disease with + test
Test −	~0 2*0.01 (false neg. error rate)	96	96	
	2	98	100	

COMMUNITY WITH A HIGHER RATE OF INFECTION

Now suppose with have a community with 20% having a disease. Here, 20 people in the hypothetical population of 100 would have disease and 80 would be disease-free. This 20% was based on a different news source that suggested that 20% was one of the highest proportions of COVID-19 in a community in the United States.

Rapid, Less Accurate Test

Let's use a rapid test with a 10% false positive error rate, 15% false negative error rate and 20% of the people tested are truly positive. With the error rates described for this test, 17 of the 20 people with disease test positive and eight of the 80 disease-free people test positive. Based on this, a person with a positive test (17 + 8 = 25) has a 68% (17/25) chance of having the disease.

	Disease Present	Disease-Free	Total	
Test +	17	8 (80*false pos. error rate)	25	17/25 = 68% chance of disease with + test
Test –	3 (20*0.15 [false neg. error rate])	72	75	
	20	80	100	

Slower, More Accurate Test

Now suppose we apply a more accurate test with a 2% false positive error rate (98% specificity) and 1% false negative error rate (99% sensitivity) to the same population. In this case, all 20 people with the disease test positive and two of the 80 disease-free people test positive. Based on this, a person with a positive test (20 + 2 = 22) has about a 90% (20/22) chance of having the disease.

	Disease Present	Disease-Free	Total	
Test +	20	~2 (80*0.02)	22	20/22 = 90% chance of disease if you have + test
Test –	~0 (20*0.01)	78	78	
	20	80	100	

HOW MUCH DOES THIS MATTER TO ME?

What do you tell Governor DeWine? What will you conclude if you test positive for COVID-19? Do you have it? If you live in a community with little disease and use a less accurate rapid test, then you only have one in six chance (16%) of having the disease. If you have a more accurate test, then you have a 50–50 chance of having the disease. Here, you might want to have a more accurate test if you test positive on the rapid, less accurate test. If you live in a community with more people who have the disease, both tests suggest you are more likely than not to have the disease. It's important to recognize that these tests are being applied in situations where additional information is available, including where people exhibit COVID-19 symptoms or where people live or work in communities with others who have tested positive.

CONSIDERING THE COVERAGE

Many a journalist has groaned as they've taken what they think is a good story to their editor only to hear, "Where's the conflict?" Conflict is among the news values which drives journalistic coverage of something – the conflict can be political, cultural, social, or, as in this case, scientific. Think of some of your favorite stories: Who are the Avengers without Thanos? Batman without Joker? The Cleveland Browns without … well … the Cleveland Browns? (Though American football fans might know the Pittsburgh Steelers and Baltimore Ravens are major rivals, all true Browns' fans know the Browns' worst enemy is themselves.) But, wait, what conflict can there possibly be in a story about COVID testing you might ask. All kinds, if you look closely, but above all there is the conflict between the first positive and then the follow-up negative tests.

The thing about conflict is that it creates a feeling of drama around something. As one marketing blog notes, "It engages us emotionally, as we get to judge the merits of the arguments, judge those who are wrong and get our righteous agreement jollies by nodding vigorously along with those we agree with"

(Zajechowski, n.d.). Not that we need more drama around COVID, but the story of testing has been complicated, particularly when it comes to the effectiveness of those tests. And a story about testing does not sound like it's going to catch a lot people's attention on its own. However, a governor in a red state who came under fire for putting in place some of the more restrictive COVID policies during the early pandemic testing positive just before a visit from the president, well, that's not a hard sell. Mike DeWine's "does he have it/does he not have it" dance provided reporters with an opportunity to explain to their audiences why some COVID tests might be more likely to produce false positives (or false negatives, for that matter), but to do so in a way that was not abstract. "In a high-profile example of a new testing frontier," reporter Sarah Mervosh (2020) wrote, "Mr. DeWine first received an antigen test, which allows for results in minutes, not days, but has been shown to be less accurate." She then goes on to say that DeWine's follow-up test used the "more standard procedure known as polymerase chain reaction," which must be processed in a lab, takes longer to process and which generally results in fewer false positives.

Mervosh also places DeWine's plight into the larger public health picture, discussing how experts think widespread testing is necessary to curb the spread of COVID and to really understand its scope, but that widespread testing in the United States has not happened yet. Antigen tests like the one that produced the false positive for DeWine had been seen as one answer to that testing issue, but "there are drawbacks: Antigen tests will miss some people who would test positive by P.C.R., with some past antigen tests missing up to half the infections they looked for" (Mervosh, 2020). A story from Cincinnati TV station WKRC did not spend too much time on the ins and outs of testing, focusing on the governor's response, he and Fran felt fine he said, although it did point out that DeWine still planned to expand testing in Ohio (WKRC, 2020). A news story from the *Associated Press* – a news agency which publishes stories on its own website but which other news outlets can also subscribe to and publish stories from – did attempt to tease about the difficulties around COVID testing in the United States. "The conflicting results," the reporters wrote of DeWine testing for positive for COVID and then negative,

> underscore the problems with both kinds of tests and are bound to spur more questions about them. Many people in the U.S. can't get lab results on the more accurate version for weeks, rather than the few hours it took the governor to find out.

(AMIRI & SEWELL, 2020)

All three stories used conflict in some way to shape their stories of DeWine's COVID tests. In the WKRC story, the conflict was just between the two test results, while both the *New York Times* story and the piece from the *Associated Press* worked in other kinds of conflicts – including conflicts among Ohioans over DeWine's restrictions and conflicts over the best approach to COVID testing. The important thing is that none of the stories allowed the conflicts to drive them into sensationalistic coverage, which is one of the criticisms of conflict-focused reporting. In these stories, the conflict created spaces for conversation and debate rather than ginned up anger that went nowhere. That kind of measured approach is necessary whenever a reporter is covering a serious story where conflict may exist if they hope to truly inform, rather than inflame (Puttnam, 2013) their audiences.

REVIEW

While the question in early August 2020 might have focused on whether Governor DeWine had COVID-19, he has been symptom-free from that date until the end of the calendar year when this chapter was first being drafted. Ultimately, the probability that the governor is disease-free reflects the chance of being disease-free given one positive result on a less accurate test and two negative results from more accurate tests. The chance that he is disease-free is very close to one.

You might be interested in controlling different kinds of errors with different tests. If you are screening for COVID-19, you might want to minimize false negative errors and to accept potentially higher false positive error rates. A false positive error means a healthy disease-free person is quarantined and unnecessarily removed from exposing others. This may be beneficial from the perspective of public health and promoting the general societal level health although an individual would not be happy to be such unnecessarily quarantined false positive. A false negative error means a person with disease is free to mix in the population and infect others.

When it comes to reporting, the kinds of conflicts that can emerge from different test results can fuel coverage, but reporters have to be careful to use the conflict to broaden the discussion of something, not to stir up sensation. This is particularly important when it comes to public health issues.

Note: Portions of this chapter appear in the blog post "My COVID-19 test is positive … do I really have it?" ISI statistician reacts to the News blog (https://blog.isi-web.org/react/2020/08/my-test-is-positive/).

STATS + STORIES PODCASTS

Stats + Stories has featured a number of conversations about uncertainty and what probabilities of outcomes mean in terms of how likely it is that an event occurs.

Alexandra Freeman and Claudia Schneider discussed how we understanding uncertainty (https://statsandstories.net/society1/how-we-understand-uncertainty). Louise Ryan focused on a similar topic (https://statsandstories.net/methods/communicating-uncertainty).

Andrew Flowers discussed how probabilities greater than 50% are not the same thing as certainties (https://statsandstories.net

/media1/2018/8/2/so-a-70-30-support-split-is-not-a-sure-thing -stats-stories-episode-29).

Nick Fisher and Dennis Trewin described building testing strategies for Australia (https://statsandstories.net/methods/a-bet ter-way-to-test-for-coronavirus).

Risk literacy and the use of natural frequencies for risk communication was the topic of a conversation with Gerd Gigerenzer (https:// statsandstories.net/education1/2018/8/23/reading-writing- and-risk-literacy-stats-stories-episode-64).

To Learn More

To read more different types of COVID-19 tests

> Coronavirus: Rapid test vs. PCR test. What you need to know. Kaitlin Schroeder 8/7/2020 Dayton Daily News https://www.msn.com/en-us/news/us/coronavirus-rapid- test-vs-pcr-test-what-you-need-to-know/ar-BB17GZTa

> How Reliable Are COVID-19 Tests? Depends Which One You Mean. Richard Harris May 1, 2020. National Public Radio. https://www.npr.org/sections/health-shots/ 2020/05/01/847368012/how-reliable-are-covid-19-tests- depends-which-one-you-mean

To read more about accuracy of COVID-19 tests

> COVID-19 Story Tip: Beware of False Negatives in Diagnostic Testing of COVID-19 – https://www.hopkins- medicine.org/news/newsroom/news-releases/covid-19 -story-tip-beware-of-false-negatives-in-diagnostic-testing -of-covid-19 (Johns Hopkins press release describing work suggesting false negative rates > 20% for RT-PCR tests and that test accuracy changes over time course of disease)

https://www.npr.org/sections/health-shots/2020/04/21
/838794281/study-raises-questions-about-false-negatives
-from-quick-covid-19-test

To read more about screening tests in a different context – facial
recognition systems

Live Facial Recognition: how good is it really? We need
clarity about the statistics
David Spiegelhalter and Kevin McConway
https://medium.com/wintoncentre/live-facial-recogni-
tion-how-good-is-it-really-we-need-clarity-about-the-
statistics-5140bd3c427d

To read more about natural frequencies in discussing screening
test risks

Gerg Gigerenzer (2014) *Risk Savvy: How to Make Good
Decisions*. Viking Press.
Do doctors understand test results? By William Kremer
BBC World Service
https://www.bbc.com/news/magazine-28166019

To read more about natural frequencies/hypothetical populations
as part of seven concepts important for being a statistically literate citizen

Jessica Utts (2003) What Educated Citizens Should
Know About Statistics and Probability, *The American
Statistician*, 57:2, 74–79, DOI: 10.1198/0003130031630

Research Reproducibility and Reporting Results

Over half of psychology studies fail reproducibility test – Largest replication study to date casts doubt on many published positive results. *Nature*; 27 August 2015 (by Monya Baker).

Scientists Replicated 100 Psychology Studies, and Few Than Half Got the Same Results – The massive project shows that reproducibility problems plague even top scientific journals (by Brian Handwerk). *Smithsonian Magazine*; August 27, 2015. https://www.smithsonianmag.com/science-nature/scientists-replicated-100-psychology-studies-and-fewer-half-got-same-results-180956426/

DOI: 10.1201/9781003023401-9

Many Psychology Findings Not as Strong as Claimed, Study Says (by Benedict Carey). *The New York Times*; August 27, 2015. https://www.nytimes.com/2015/08/28 /science/many-social-science-findings-not-as-strong-as -claimed-study-says.html

Research Publication:

Open Science Collaboration (2015) Estimating the reproducibility of psychological science. *Science* **349**: aac4716-1–aac4716-8.

Pixabay from Pexels (https://www.pexels.com/photo/pile-of-covered -books-159751/)

STORY SUMMARY

Major scientific and medical journals often issue news releases to highlight research they believe will capture the attention of the public. For studies viewed as exceptionally exciting, journals may impose an embargo, which means a news outlet cannot publish a story about a study until the journal issue it appears in has been published. Embargoed or not, press releases are translated into headlines designed to catch the attention of a reader in the hope they will then consume the entire story. But an important question remains: Is the result reported in the story real? Would others conducting the same study in a similar population or in a different population obtain the same result? A series of headlines in high-profile news outlets (*New York Times*), general audience publications (*Smithsonian Magazine*) and scientific outlets (*Nature*) described the results of an extensive study which sought to understand whether the results observed in a specific set of scientific studies could be replicated. The short answer: No. The headlines of all those publications suggest an inability of repeated experiments to confirm the results of the initial studies in question and, consequently, cast suspicion on published positive study results.

The *Science* study that provided the foundation of these stories focused on 98 unique psychology experiments that had originally been published in three top psychology journals: *Psychological Science* (PSCI), *Journal of Personality and Social Psychology* (JPSP), and the *Journal of Experimental Psychology: Learning, Memory, and Cognition* (JEP:LMC). A "top" journal is one that is viewed within a scientific discipline as the most desirable and prestigious outlet for research in a discipline. It will often be the home of novel, exciting and even surprising research results. The papers published in these top journals are often among the most cited studies in a particular discipline and frequently serve as the foundation of stories in the popular press.

In the *Science* paper we're discussing, a couple of studies were replicated by more than one lab, resulting in a total of 100 attempted replications. One remarkable aspect of the *Science* study was that a large, international team of scientists, part of the Reproducibility Project, conducted this massive task. What all these attempts to replicate the original research found is that fewer than half of the repeated studies obtained the same result as the original study. One aspect of the failure to obtain the same result was quantified in terms of not obtaining a similar result, described in terms of "significance – a statistical measure of how likely it is that a result did not occur by chance" New York Times (NYT) or "whether a statistically significant result could be found" (*Nature*).

What is so interesting in the 2015 coverage of reproducibility in relation to psychology studies – or its lack thereof – is its explicit challenge of the practice of rushing to celebrate a novel result that may need to be replicated before being embraced as some truth about the world. It provides caution and a call for critical reading of claims in the news.

WHAT IDEAS WILL YOU ENCOUNTER IN THIS CHAPTER?

- Confidence in study/experimental results accrues with the replication of findings.

- Replication may involve repeating a study using identical conditions, to the fullest extent possible, possibly with different populations.

- Studies that are published are a subset of all studies and may reflect a surprising result.

- Scope of impact can drive news coverage.

- Duty of care in relation to journalism.

WHAT IS CLAIMED? IS IT APPROPRIATE?

Repeating the same experiment will more likely than not fail to realize the same outcome. This assertion was based on a study of 98 experiments, selected from papers published in top psychology journals. Here, these studies were repeated by a collection of independent researchers. This claim is captured in terms of a dichotomy of whether the second repeated experiment provided the same conclusion as the original statistical test. In addition, the replication study reported that the size of the impact or effect of the experiment observed in the repeated experiment was generally smaller than that observed in the original experiment.

WHO IS CLAIMING THIS?

The journal *Science* is published by the American Association for the Advancement of Science (AAAS) – a professional society with more than 100,000 members. (If you are interested, check out the bibliometric measures of the impact of this journal. It is clearly no slouch in the world of scientific research publications.) Here, a "top" journal (*Science*) is publishing a study of whether projects published in "top" (psychology) journals can be replicated. Would a possible next project be a replication study of replication studies?

A consortium of researchers, the Open Science Collaboration (OSC)/Reproducibility Project: Psychology (https://osf.io/ezcuj/) were the authors of the paper "Estimating the reproducibility of psychological science."

The OSC self-description states that it is "[a]n open collaboration of scientists to increase the alignment between scientific values and scientific practices." Based on this published paper, we can infer that an important scientific value is that reported scientific discoveries should be confirmed by experiments that follow the same protocols. Further, the scientific practice is implicitly the underlying process by which papers move from manuscripts to peer-reviewed publications.

WHY IS IT CLAIMED? WHAT MAKES THIS A STORY WORTH TELLING?

The latest issues of highly reputed scientific journals often include papers that are featured in press releases and picked up by journalists. Often the more exciting and novel results are the focus and feature of stories. This particular scientific paper and the subsequent coverage in high visibility outlets provide confirmation of this pattern. In particular, the declaration that fewer than half of the experiments have been replicated casts a serious cloud over the overly enthusiastic and uncritical promotion of research studies. Here, the scientific paper that is the basis of the news coverage is a meta-study of the reproducibility of other scientific papers.

The story that fewer than half of the repeated experiments got the same results is based on a number of analyses that included measures associated with statistical hypothesis testing (P values and compared to stated levels of significance), effect sizes and meta-analysis, the methods that pool study-specific results such as estimates of treatment effects to obtain an overall estimate of the impact of some treatment. In this chapter, we focus on the first measures.

This story is worth telling because of the concerns over the potential for a reproducibility crisis associated with scientific research. If science is predicated upon the ability to reproduce results and then building new research upon that tested and re-tested foundation, an inability to reproduce findings creates a foundation with the potential to collapse. It can also serve as a cautionary tale about not pinning too much on the results of a single study.

IS THIS A GOOD MEASURE OF IMPACT?

Do you agree with me or not? What could be a simpler question to ask? For each of the 98 previously published studies, the count of how often a repeated experiment agreed with the original experiment is one criterion for evaluating reproducibility. Another criterion would be the direction and size of change associated with some treatment or association among variables being studied.

HOW IS THE CLAIM SUPPORTED?

The claim is supported by repeating the original studies and comparing the original study results with the replication study results. This is quantified in a variety of ways including a comparison of the results of statistical hypothesis tests along with other methods.

What Evidence Is Reported?

You might think that reproducibility is a simple exercise where the yes/no answer to the question of a significant result in the original study is paired with a yes/no answer to the same question in the replication study. So one might imagine the following table of results:

		Replication Study	
		Significant	Not Significant
Original study	Significant	Hurray! ☺	Rats ☹
	Not significant	Rats ☹	Hurray! ☺

Where the diagonal entries ("Hurray!" cells) correspond to the situation where the replication studies echo the original study's qualitative conclusions, the off-diagonals ("Rats" cells) correspond to the situation where there is a disagreement between the original and replication study results. If all results were in the diagonal, then there wouldn't have been a story – or perhaps it might be even a bigger story! If the total agreement was the result, you could change the headlines:

- All psychology studies pass the reproducibility test – the largest replication study to date affirms all published positive results.

- Scientists Replicated 100 Psychology Studies, and All Got the Same Results – the massive project shows that there is no reproducibility problem.

- Psychology's fears unfounded: rechecked studies hold up.

- Only need to do one study to establish the truth of scientific observations in psychology!

Would these headlines have attracted readers? Maybe. The last hypothetical headline is the scariest of the set in our eyes as it would suggest that science requires only one study to establish some foundation of truth. Actually, if all had been replicated, then another possibility for a headline might be "Widespread Collusion in Psychology Research Circles" since we doubt anyone would believe such a report.

We'll revisit the question of "significance" soon, but first let's think more about what studies were available. The original studies already represent a special subset of research work: they were published in top journals. Acceptance rates for top academic journals can be quite low, particularly for high-impact scientific journals, which means there's a lot of work that never finds its way into a journal. We don't know anything about papers that were submitted to these journals but weren't accepted. Finally, novel and exciting results may have a better chance of being published than a study viewed as more mundane, such as the replication of a previously published study.

Would scientists who didn't have a significant result submit their paper to a journal? *We did the following experiment and nothing exciting was observed.* (As one of our graduate school professors once told us: "Zero is a number, but not a very interesting one.") What might happen if the authors decided to submit the paper that didn't have a significant result? Journal referee 2 report: *The authors conducted a reasonable experiment but they didn't see anything significant and interesting so I recommend rejecting this paper.*

Considering what even permits a study to be in the pool of studies that would be candidates for replication, you might imagine that the following table better reflects the reality of the situations being explored.

| | | Repeated Experiment | | |
		Replicated (Statistical Significance and Same Direction)	Not Replicated	Total
	Published ('significant')	👍 Can answer	👍 Can answer	96
Original experiment/ research	Published ('not significant')	👍 Can answer	👍 Can answer	4
	Submitted but not published	?	?	?
	Conducted study but not submitted	?	?	?

The take-away message from this table is that not all research is submitted, not all submitted research is published, and it is a rarefied set of research results that are finally published in top-tier scientific journals. Thus, our question of "reproducibility" is framed relative to results and studies that are published in top journals in the field. This will reflect the quality of the experimental methods and the interest in the results of the work.

Ok, we're now ready to consider what "statistical significance and same direction" means.

What Is the Quality/Strength of the Evidence?

Statistical hypothesis tests start with competing beliefs about nature, with one of these beliefs being that some difference between groups or relationship between variables exists (usually the reason the study was initiated in the first place!) versus there is no difference between groups or no association between characteristics or that experimental manipulations do not change a response. (We talk about this process in our chapter about Nike's reportedly swift running shoes.) Data are collected and some function of this data (a test statistic) is calculated. Often a P value is then constructed.

If there is no difference or effect present, then the P value is the probability of seeing a result at least as extreme as observed. So basically a small P value suggests that either a very rare outcome was observed when truly there is no difference (no treatment effect) *or* there is really a difference present. Now, embedded in this calculation are the assumptions that are the foundation of the test statistic, but let's assume these are correct. So how small is small enough? One common level of "sufficient" surprise is 1 in 20 or 0.05, a pre-declared significance level. If the P value is less than or equal to 0.05, then a declaration of statistical significance is made. Note that this doesn't mean it is an important difference. Really large experiments often will declare statistical "significance" even when the size of the difference is quite small.

As a simple example, consider a series of simple studies where a fair, balanced coin is flipped and the number of heads is observed. If you don't like tossing two-sided coins, an equivalent study is rolling a balanced six-sided die and observing 1, 3 or 5 pips or drawing a red card from a well-shuffled 52-card deck.

Number of Tosses	Number of Heads Observed	Proportion of Tosses Resulting in "Heads"	P Value for a Test That the Coin is Balanced	Confidence Interval (95%) for the Proportion of Times the Coin Toss Results in a Head
100	51	0.51	0.920	(0.4087, 0.6106)
1000	510	0.51	0.548	(0.4785, 0.5414)
10,000	5100	0.51	0.04659	(0.5002, 0.5198)
100,000	51,000	0.51	0.0000000002592	(0.5060, 0.5131)

While the number of tosses ranges from 100 to 100,000, the number of heads observed corresponds to a 51% (0.51 proportion) of heads being observed. If you use a test of whether the true proportion of heads being observed in ½, you obtain P values that decrease with increasing numbers of tosses with 51% heads in 100,000 tosses being associated with a P value of much smaller than 1 in a million. From a formal hypothesis perspective, the 51,000 heads in 100,000 tosses provide evidence that the coin is not balanced – has a probability of heads that differs from 0.50. A natural question is to ask how far from ½ does the estimate fall. In this 100,000 coin toss study, there is evidence that the coin toss probability of heads ranges between 0.6% and 1.3% larger than 50% and between 0.02% and 1.98% larger than 50% in the 10,000 coin toss study. Is this an important difference from 50%? This effect size is often a more interesting question than some P-value comparison.

Now, back to the *Science* paper. Here, P values were available from the original research papers *and* from the repeated experiments. In addition, the direction of the effect was considered. Here, if the experiment originally resulted in a larger response in one group versus another, a common direction of effect was observed in the replicate study if the larger response was observed in the same group relative to another.

If you look at original studies (P value below 0.05 significance level – almost all of them) and at replicate studies (P value below 0.05 significance level), you will see 35 of 97 (36%) replicate studies resulted in the same qualitative declaration of statistical significance. This replicate proportion varied among the different journals from 23% (7/31; JPSP, social) to 53% (8/15; PSCI, cognitive). What does this mean? At a minimum, the findings of the original study are not reproducible.

DOES THE CLAIM SEEM REASONABLE?

The claim that (psychology) studies are not reproducible seems surprising. From a default perspective, we expect that a published scientific paper has been vetted by a journal and will reflect some truth. It is easy to forget that we are studying a sample from a population and that we expect samples will differ. In addition, the different studies may be from different populations. The population of 18- to 22-year-old adults may differ between one comprising adults attending university and another comprising adults who entered the workforce following secondary school. There may be universal truths in the research applicable to both groups or there may not. Given the systematic effort invested by a large collection of scientists to reproduce study results, the claim is believable, just somewhat surprising.

HOW DOES THIS CLAIM FIT WITH WHAT IS ALREADY KNOWN?

The *Science* paper, along with the headlines and stories associated with this paper, all mention concerns about replication in science and some mentioned that analogous efforts to replicate experimental results have emerged in other disciplines, including cancer biology. Scientific conferences have organized panels around the possibility of a reproducibility crisis in science and science reporters have been writing about the issue as well. The fact that the paper in *Science* could not reproduce the results of studies in a particular field is a major contribution to this debate.

HOW MUCH DOES THIS MATTER?

Does science matter? Does the systematic study of our world matter? Are our beliefs about the world shaped by scientific inquiry? YES × 3! This harkens back to early exposure to ideas from the scientific method where we are taught that science advances from the accumulation of knowledge and that scientific theories can be supported by studies, or possibly rejected by studies. From that perspective, the *Science* study was an important reminder about experimental support for theoretical claims needing to accumulate over time.

Comparison of Population Perspective versus Individual Perspective?

It is interesting to think about what constitutes an individual and what constitutes a population here. It may feel strange to think about a single study as an individual from some population of studies of a phenomenon, but perhaps that is a good way to start. If you believe an assertion becomes more compelling as evidence accumulates, then you have a great start for placing a single study in context. Repeating an experiment or study with the same conditions in the same population or in different populations or even with somewhat altered conditions is analogous to increasing the evidence base.

Will I Change My Behavior as a Consequence of This?

We don't know about you, but this issue of reproducibility does give us pause, particularly when it comes to the circulation of news stories and journal publications on social media. It can be too easy to retweet or reshare a news headline that makes a splashy claim or a journal article that, on its surface, looks set to remake a particular field – but we should always be looking for the broader context. That doesn't necessarily mean you *shouldn't* share a news story about a new study, or even the study itself; what we do think it means is that we all would be better served if all of us – reporters, academics and the public – took a deeper dive into the information before sharing it. One thing reporters and researchers alike could do is dive into the original research paper (not just the news release) and see if an experimental result has been observed in other populations or by other researchers. Are there mechanisms that would explain why this result occurred? It might be cliched, but there's a reason moms and dads tell their kids to be careful of things that look too good to be true – they often are. In journalism and academic writing, falling too hard for a novel result not only can lead to the circulation of untested ideas but can also erode your own credibility as a reporter or scholar.

CONSIDERING THE COVERAGE

Duty of care is a legal concept, one that suggests that individuals and businesses have a duty to take all reasonable precautions to minimize harm during their interactions with others. British filmmaker David Puttnam gave a TED Talk in 2013 in which he suggested news media also have a duty of care they should strive to live up to in the work they do. Reporters have to choose, he said, whether they seek to "inflame or inform" the audiences they speak to. Puttnam was talking about news media coverage of political issues – but one might ask the same question of reporters writing about scientific discoveries. Do reporters want to inflame or inform their audiences? The inflammatory path would be the sensational one – writing breathlessly about a discovery or finding without providing context or nuance. The informatory approach would be cautious in its celebration, choosing to focus on what the finding adds to our understanding while also addressing the need for the finding/s to be confirmed by other researchers before we hold too tightly to them. It can be difficult to know, as a reader, the path a reporter has chosen; however, one thing we can all do as we read news stories (about anything) is check our emotional response to them. Does a story make you immediately angry? Does it cause you to feel disgusted? Do you finish a news story enraged or frustrated? If so – ask yourself why. There are reasons a news story might upset you, and your feelings might be related to the background or the context a reporter provided. But, if you have an immediate and visceral response to a story, and there's not a lot of detail to be found in it, that might suggest that the story is pitched to inflame. Inflammatory stories are often heavy on emotion, but light on details.

The reporting on the *Science* study about the reproducibility of psychology studies attempts to contextualize the study's findings, although the *New York Times*'s opening sentence "The past several years have been bruising ones for the credibility of the social sciences" (Carey, 2015), perhaps leans a bit too far into the inflame side of things. A few paragraphs into the story the reporter also writes that "Their conclusions … have confirmed the worst fears of scientists who have long worried that the field needed a strong correction" (Carey, 2015). *Nature* is similarly blunt in its opening paragraph about the study, the writer suggests "Don't trust everything you read in the psychology literature. In fact, two thirds of it should probably be distrusted" (Baker, 2015).

While the study certainly adds support for what some have called the "reproducibility crisis" facing some scientific fields, the reporting from the *New York Times* doesn't explain that, troubling though some may find the *Science* study's results, this is actually the scientific process at work. That's something Brian Handwerk, writing for *Smithsonian Magazine,* does point out.

> The eye-opening results don't necessarily mean that those original findings were incorrect or that the scientific process is flawed. When one study finds an effect that a second study can't replicate, there are several possible reasons … Study A's result may be false, or Study B's results may be false – or there may be some subtle differences in the way the two studies were conducted that impacted the results.
>
> (HANDWERK, 2015)

Looking back on the study a few years later, *Slate*'s Daniel Enberger (2017) ponders whether the study's findings suggest a true crisis or are simply evidence of science self-correcting. He then goes on to write about how the study's findings were politicized by some who are looking for any reason to discredit scientific research. "With that," Enberger wrote, "another of scientists' biggest fears was confirmed: that any discovery of major problems in their field would end up being used against them." In essence, scientists' honesty about what they found was used by some to undermine trust in the institution as a whole. And journalists, by potentially catastrophizing the problem (Enberger 2017), produced stories that "served as chum for anti-science trolls." Enberger, however, does suggest there might be something "broken" in the scientific process.

> In the last few years we've learned that science sometimes fails to work the way it should. Suggesting it might be "broken" is not the same as saying it's in a state of utter, irreversible decrepitude—that every published finding is a lie, or that every field of research is in crisis.
>
> (ENBERGER, 2017)

That's an important point for us as readers to remember as well – that even if the scientific process is messy at times, or even broken, it's not untrustworthy. Nor, does messiness in one part of science throw all of science, and the scientific process, into question. Being a critical reader is important, and always has been, but it doesn't mean assuming everything is a "lie" as Enberger puts it. We live in a sea of information – to be an ethical reader in such a situation, and to minimize harm as readers, demands that we do not let skepticism blind us to uncomfortable truths even as we work to more critically engage with the information we consume.

The question becomes how to cover this state of affairs as a journalist ethically? How can you inform an audience and avoid inflaming it? Reporters, like researchers, have no control over how their output is taken up and used by others, which is where the idea of a duty of care can be so useful. All duty of care asks is that you take "reasonable precautions" to minimize harm. Minimizing harm is, after all, one of the ethical mandates championed in the Society of Professional Journalists' code of ethics. What does it mean to "minimize harm" in relation to science reporting?

In our view, minimizing harm can be accomplished by providing context for what you report and by not rushing to celebrate, or debunk, something based on one study. Perhaps the most important take-home message is that a single study is never the definitive, final answer to a scientific claim, and expectations by all of us need to be tempered by this realization. If journalists cover science stories, even ones like the reproducibility crisis, with care to avoid invoking an inflammatory response, then all they can do is publish or broadcast the piece and hope for the best. But, exercising a duty of care by providing context for stories will help ensure they aren't automatically turned into chum to feed a circling sea of trolling sharks.

REVIEW

Research and scientific investigation provide one way we learn about our world. Confidence in the validity of scientific research increases with replication. If the results from a research study can't be replicated, then does this cast doubt on the work? Stories associated with an extensive reproducibility study in psychology were explored along with the reproducibility study that provides the headlines for the stories. Evaluating agreement between original studies and subsequent, replication trials provided important information. Studies published in science represent a subset of research that is submitted which is a smaller subset of research that is conducted. For original studies published and included in the replication meta-study, measures of agreement between the original study and the replication attempt are constructed. Agreement in dichotomous declarations of statistical hypothesis testing results based on *P* value comparison to a fixed level is then tabulated. The discordance between the declarations of the original and replication studies led to the headlines. This was not an invalidation of science but, arguably, reinforces the importance of the process and provides a caution about overenthusiasm for study results.

Ultimately, there are many possible reasons that a study might not replicate. It might not replicate because the first study reported an effect that wasn't true – a false positive result. It might not replicate because the second study employed a sample of observations from a different population or employed a protocol that differed in some way from the initial study. From the perspective of a general reader, what do you notice about the population studied? In an earlier chapter, we explored the differences in shoes among elite runners. Would this study be replicated in a different population? Could it?

When it comes to reporting, the duty of care is a framework that can help reporters slow down a bit when it comes to the coverage of novel research results. It asks that reporters take "reasonable" precautions to minimize harm during the reporting process. Stories about scientific research often drive interest in science and the more surprising the story, the more excitement that's stirred up in readers. But that excitement can sometimes be a mixed bag. Take for instance early enthusiasm for the discovery of pentaquarks – a subatomic particle made of four quarks and one antiquark. A lab in 2003 claimed to have discovered their existence, with journalists – including one of this book's co-authors – writing stories lauding the discovery. However, scientists tried for years to replicate the work – with some labs claiming to have done so, while others could not. There was also the issue of whether the data proving the existence of pentaquarks was sound. The back-and-forth called into question the early coverage of the particle's discovery. And then, in 2015, the Large Hadron Collider found evidence that pentaquarks do, in fact, exist. It took 12 years for the saga to play out and, while those original findings were supported, the pentaquark saga should serve as another reminder of the importance we should attach to the reproducibility of surprising scientific findings. In relation to news coverage of scientific discoveries – whether they are focused on psychology or pentaquarks – news media would serve society, and their audiences, better if they communicated more clearly how science operates as an iterative process where knowledge and understanding accrue over time. A single study result must be placed in context and enthusiasm tempered by that context. Replication rarely provides a clickbait tempting headline, but perhaps it should.

CODA: A NEW 3 R'S?

Many of us learned the 3 R's of Reading, wRiting and aRithmetic (a stretch but it almost works). We may have a new set of R's to learn – repeatability, replicability and reproducibility. Hans Plesser reflected on some of these new R's in a short note in 2018, suggesting that there has been some confusion over the use of terms. His note came a few years after the Association for Computing Machinery outlined its own approach to these new 3 R's, which is paraphrased below as:

- Repeatability – same research team using the same experimental set-up obtaining similar results.

- Replicability – different teams using the same experimental set-up obtaining similar results.

- Reproducibility – different teams using a different experimental set-up obtaining similar results.

The concern about reproducing/replicating studies has led National Academies of Science to author a report about the issue in 2019. In it, **reproducibility** emphasized computation – the same results obtained from computing with the same input, steps in coding and computing and related issues – and **replicability** emphasized consistency of results from different studies. In addition to the 3 *R*'s, the idea of **generalizability** is an important addition to the conversation at it captures the idea of whether a result in one study is observed in a study using a different situation, such as a different population. What are readers – statisticians, journalists or the general public – to make of all of this? We think the last sentence from the summary of the National Academies of Science report gives us this useful reminder that "reviews of cumulative evidence on a subject, to assess both the overall effect size and generalizability, is often a more useful way to gain confidence in the state of scientific knowledge." To return to our metaphor of earlier, it suggests we should all focus more on how things come together to form a foundation, rather than a single element of that foundation when reading scientific studies or news stories about them.

STATS + STORIES PODCASTS

Stats + Stories has featured a number of conversations about the statistical hypothesis testing and the *P*-value report of American Statistical Association (ASA) and hypothesis testing.

These include Robert Mathews in the episode: To *P*, or Not to *P*, That is the Question | Stats + Stories Episode 194 /July 8, 2021 (https://statsandstories.net/methods/to-p-or-not-to-p-that-is-the-question) and Nicole Lazar directly addressed the ASA report on Reevaluating *P* Values | Stats + Stories Episode 93 / April 25, 2019 (https://statsandstories.net/methods/reevaluating-p-values).

David Spiegelhalter discussed research study quality in the episode: I'd Give That Study 4 Stars: Considering the Quality Of Research | Stats + Stories Episode 27 /February 28, 2017 (https://statsandstories.net/media1/2018/8/2/id-give-that-study-4-stars-considering-the-quality-of-research-stats-stories-episode-27).

CHAPTER 10

Now, What?

"Statistical thinking will one day be as necessary a qualification for efficient citizenship as the ability to read or write."
– Samuel Wilks paraphrasing H.G. Wells.

Over the last several years, scientific journals have begun a purge of articles based on racist or sexist measurements or conceptualizations (Marcus & Oransky, 2020). Just as 2020 was coming to a close, the journal *Psychological Reports* retracted two articles "following a review that found that the research was unethical, scientifically flawed, and based on racist ideas and agenda" (Psychological Reports, 2020). Throughout this book, we have discussed different statistical, measurement and journalism concepts you might need to understand in order to peel back the story you encounter in the news. What we've largely left alone – until now – are the societal implications of statistics and their interpretation, not because they are unimportant, but simply to try to communicate the fundamentals as clearly as we could first. But, statistical concepts are not mere abstractions; they are used to measure real-world phenomena and can have real-world implications. The population counts done as part of the U.S. Census, for example, are tied to representation in Congress and to the allocation of billions

DOI: 10.1201/9781003023401-10

of dollars of block grants to communities. Recorded births and deaths are used to measure the health of a community. Statistical methods to analyze experiments such as clinical trials provide the assurance about the efficacy and safety of new treatments (or vaccines as was the focus of much interest in late 2020 and early 2021).

The questions to continually ask oneself include:

- What question is being addressed with data?

- What data and variables are available?

- How are these variables measured? How do I know that this is measuring what I want it to measure?

- What analyses are being conducted to extract meaning and insight from these data?

A sibling question must be: How do I ensure social, cultural, political and personal biases do not influence the measures I choose? Or, the way I conceptualize research questions? But these are not just questions to consider as you encounter statistical concepts, they are also questions to consider as you encounter reporting about them in news stories.

Bias is something that journalists also try to work to mitigate in their reporting. We are, as humans, inherently biased to view the world in particular ways. That is neither good nor bad, it is simply a product of the place, time and culture in which we grew up. Ideally, journalists work to mitigate bias in their reporting by approaching their stories in as objective a manner as possible. The important thing to remember as you consume news content is that straight news reporting does not push an opinion – opinion resides in the editorial section of news outlets. It's important to remember, because opinion pieces do not have the same expectation to mitigate bias. While there are numerous ways to conceptualize journalistic objectivity, scholar Michael Bugeja has been quoted as saying journalistic "Objectivity is seeing the world as it is, not how you wish it were" (Cunningham, 2003). Values are often central to ideas about objectivity in the scientific community – or, at least, the idea that values should *not* be central. Rykiel (2001) notes that "Scientists typically portray the information they provide to the public as objective and value free, with the implication that those traits confer greater weight to their opinions than should be accorded to the value laden opinions of non-scientists" (p. 434). Rykiel then goes on to remind the reader that the idea that scientific judgments are "value free" is often disputed in the scientific community, particularly by individuals who see it as their duty to use the knowledge they help uncover and/or create to work toward the common good although "common good" may not always be a consensus opinion in society. It is important to note that scientists including statisticians also are products of the place, time and culture in which they live and work. While their work is tied to the assessment of hypotheses about nature that might be supported or rejected by data and analyses, it is an enterprise that is conducted in context.

What does all of this mean and why should you care? Well, for starters, no matter how much we try to work toward neutrality in what we do – whether it is statistical or journalistic in nature – we can never, truly be neutral. Relying on our methods alone to create an edifice of neutrality is meaningless if those methods have been shaped by racist, sexist, classist or some other kind of problematic understanding of the world. This is something Angela Saini has written about her in book *Superior*, as well as in other publications. For instance, Saini writes that race has been used to explain seeming differences in intelligence between people.

> The belief that races have natural genetic propensities runs deep. One modern stereotype, for instance, is that of superior Asian cognitive ability. Race researchers, including Richard Lynn and the late John Philippe Rushton, have looked at academic test results in the

United States and speculated that the smartest people in the world must be the Chinese, Japanese and other east Asians. When intelligence researcher James Flynn investigated the claim for work he published in 1991, he found that in fact they had the same average IQ as white Americans. Nevertheless, Asian Americans tended to score significantly higher on SAT college admission tests. They were also more likely to end up in professional, managerial and technical jobs. The edge they had was therefore a cultural one – more supportive parents or a stronger work ethic, maybe – endowed by their upbringing. They simply tended on average to work harder.

(SAINI, 2019)

Later in the same piece, Saini quotes historian Evelyn Hammonds as saying, "I think that scientists, they are trapped by the categories they use. They will either have to jettison it or find different ways of talking about this... They'll have to come to terms with that it has a social meaning" (Saini, 2019). It is the social meaning of statistics, of their conceptualizations, utilizations and interpretations that we want to urge you to keep in mind as you consume news media about research, as you read research, or, if you are involved in research, as you conduct your own studies, and if you are involved in reporting the news, as you write stories based on research. This is not just a numbers game, but it's one of meaning as well. What does a particular set of numbers mean beyond whether one variable is associated with another, particularly if there is an assertion that one variable is causing changes in another? Why does it matter that something might have predictive value? Do these numbers actually measure what they say they're trying to measure?

In this book about statistical concepts and journalism about them, we hoped to provide common anchors for reflecting on these concepts. Throughout the book, we framed our review of news in terms of eight questions:

1. What is claimed and is it appropriate?

2. Who is claiming this?

3. Why is it claimed?

4. Is this a good measure of impact?

5. How is the claim supported?

   - What evidence is reported?

   - What is the quality/strength of the evidence?

6. Does the claim seem reasonable?

7. How does this claim fit with what is already known?

8. How much does this matter for me?

   - Comparison of population perspective vs. individual perspective?

   - Will I change my behavior as a consequence of this?

Here, we supplement this set of questions with a list of strategies to help you navigate the stats and the stories you encounter in the future.

CONSIDER THE WEIGHT OF EVIDENCE

For decades, climate scientists were urging politicians and the public to take what was then called global warming seriously. While there were debates in the scientific community about the specific mechanisms that seemed to be speeding up global warming, the scientific consensus was that humans were causing the escalation. However, there were outliers who claimed that was not the case, that the warming globe was simply the result of the earth's natural cycle from cool to hot to cool again. The weight of the evidence for climate change, as we now know the phenomenon, supported the scientific consensus. Those outliers, though, were able to use ideas about fairness and objectivity to have an outsized influence on the tenor and tone of the public debate. This is not to say that an outlier perspective may not be correct on occasion. For instance, Galileo's heliocentric theory of the solar system ran counter to the geocentric popular opinion but subsequent data and analyses provided support for Galileo's ideas.

Journalism, at times, gets trapped by its desire to appear objective. This can produce a **both-siderism** that amplifies fringe perspectives … or scientists. In the case of climate science, because there was a tiny minority of researchers who claimed there was no evidence that global warming was man-made, journalists felt compelled to give them space in their reporting. This made it seem like, to the casual news consumer, that the scientific consensus on global warming was on much shakier ground than it actually was. Communication research has shown that public opinion often sways political opinion on issues; had journalists in the United States reported in a more sound way, one that was reflective of the actual weight of evidence on climate change, perhaps politicians would have been pressured to take action on the problem decades ago. Instead, current debates often center on whether we can mitigate the impact of climate change – how much of Florida are we willing to cede to a rising ocean – not avoid it. The weight of evidence might have steered journalists and the public right in another situation as well.

In 1998 the medical journal *Lancet* published an article that claimed to have found evidence of a link between the MMR vaccine (which prevents measles among other illnesses) and autism. In 2010, the journal retracted the article after it was found that a number of things in the piece were "incorrect" (Eggerston, 2010). In the 12 years between publication and retraction, however, the article wreaked havoc on public health. There have always existed vaccine skeptics, and there may be real reasons for people to be uncomfortable with the standard childhood vaccination schedule, but vaccines save lives. The measles, mumps, and rubella (MMR) vaccine in question has virtually wiped out measles in many communities. However, people in what has been labeled the anti-vaxxer movement held up the *Lancet* article as evidence that there was a medical foundation for their concerns over the use of that vaccine. Journalists covering the so-called vaccine controversy treated the anti-vaxxer movement's arguments with the same care they treated those of vaccine scientists. Celebrities with no scientific training were being interviewed on news programs to talk about the dangers of vaccines even as other scholars could not reproduce the findings of the *Lancet* study. Journalists claimed they were attempting to be objective in their reporting when, had they really examined the weight of the evidence, they would have seen there was a very brittle foundation for the anti-vaxxers' claims. If objectivity is truly "seeing the world how it is," then journalists missed the mark when it came to the reported connection between autism and the MMR vaccines, with real-world consequences. While this reflects yielding to a both-siderism temptation, other issues might be at play. News and journalism are competing for attention among an ever-growing collection of news sources that spans traditional news organizations, social media outlets and more. Attention to a story and clicks translates into advertising support. What generates attention? Conflict and controversy. Stating that all but a few scientists believe something doesn't produce the energy that conflicting opinion might.

The Centers for Disease Control and Prevention (CDC) reports that in 2019, just before the global outbreak of COVID-19, there were 1282 reported cases of measles in the United States (CDC, n.d.). That's the highest number of recorded cases in the country since 1992. The CDC notes two possible reasons for the spike: people traveling outside the United States and "the spread of measles in U.S. communities with pockets of unvaccinated people" (CDC, n.d.).

CONSIDER THE SOURCE

The source of research results matters. In our chapter review questions, we asked who is claiming the results that were reported and why the claim is being made. Research is often funded by some entity and good journalism will ensure the funding for a project is made clear. Some journals are associated with professional scientific societies and have extensive and careful review processes. Others are not. This is not to suggest that if you follow the source you will always find a reason to discount – or celebrate – a study's findings; however, understanding who has funded a study or what groups are supporting the work of a researcher might help you understand why a particular focus has been taken or methodology adopted over another. While journal articles generally have spaces where any funders or conflicts of interest are acknowledged, not every news story reports a study's funder. Often that's simply an oversight, but if you don't see that information in a news story, it's not a bad idea to visit the original journal article to see if there is any mention of a funder there.

CONSIDER THE HISTORY

Race has become a standard demographic when it comes to measures of public health, poverty and access to housing. During the COVID-19 pandemic, it quickly became clear that communities of color were suffering at disproportionate rates (The COVID Tracking Project, n.d.). But is it really the race of these individuals that is making them sicker? Or is it something else? There's certainly a preponderance of evidence that black mothers and babies do less well than their white counterparts in the American healthcare system. But, again, the question is whether that is because of their race or if race is actually an indicator of something else. As noted earlier, race is a variable that is measured as part of some analyses. What does this represent? Is this simply a social construct that serves as a surrogate for other variables in an analysis?

In the case of the health of black mothers and babies, structural racism, not race, has been identified as a major factor in their health (Taylor, 2020; Wallace et al., 2017). Scholars Deidre Cooper Owens and Sharla Fett (2019) note "Since the 1990s, research on maternal and infant death disparities has increasingly pointed to structural racism in society at large as a stressor that harms African American women at both physiological and genetic levels" (p. 1343). Racism has shaped the medical community's understanding of black bodies, the authors argue, which has led to adverse impacts in the African American community. The question when it comes to differences in health outcomes comes down to, again, whether it's race or racism.

This question also comes into play in research into income inequality and access to affordable housing. On average, black Americans make less money than white Americans and are less likely to own their homes (Conley, 2010). If we just consider the numbers – gathered during the U.S. Census Bureau's regular counts – the fact of racial disparities in terms of wealth becomes clear (Chetty et al., 2020). But, if we dig into the history of those numbers, we might uncover redlining in particular communities that kept black families out or the way that slavery and segregation impacted the ability of African Americans to amass wealth. The numbers give us an idea that something may be going on, but it's the history that sits behind those numbers that really helps us begin to develop a nuanced picture of what's happening.

BE A CRITICAL READER ... OF EVERYTHING

A news story and the research behind it should never be the end of the story – it should open up new questions and new paths of inquiry. Tim Harford (2021) suggests that being curious is the most important trait for a person who wishes to be a data detective, and we agree. If there's a preponderance of evidence supporting something, it's worth asking how that evidence was amassed and whether reporters, researchers and scholars have overlooked something. Journalists are constantly looking for alternate interpretations for things they cover; statisticians also keep in mind there are other ways of interpreting data and whether there are missing variables that might be important for understanding some phenomenon. If those alternate interpretations don't seem sound or there is no evidence to support them, then you can feel comfortable that the story your stats are telling you, or the stat you're using in your story, is solid. Look to see how this story is being covered by other news sources. Other sources are a browser search away.

To be a critical reader is to question everything, but not in a pedantic way; rather, the questioning should help you better understand the choices a scientist made when conducting, analyzing and presenting their research as well as the choices a journalist made when covering a study. To be a critical reader is to be aware and observant. You understand statistical concepts and have a basic understanding of how journalism works, all that's left is for you to put this knowledge into practice as you design your own studies or as you read news coverage of them. It is our hope that this book – with its discussion of the statistical concepts needed to understand research and the journalistic concepts needed to understand the news coverage – will help you approach both research and journalism with a critical eye. This is ever more important as data show up in almost every aspect of our lives. It certainly feels like we are living in that moment H.G. Wells predicted, where being quantitatively literate is as important as being textually literate. Understanding how data is produced and how it is reported in news media can help all of us make more informed decisions about not only the work we do, but about the choices we make in our everyday lives.

Bibliography

Amiri, F., & Sewell, D. (2020). DeWine tests negative after positive before Trump visit. *Associated Press*, August 6. https://apnews .com/article/columbus-ap-top-news-mike-dewine-politics-virus -outbreak-e4750f1deb1bf375d968e33dc6ff9051

Arnold, J. B. (2019). ggthemes: Extra themes, scales and geoms for 'ggplot2'. R package version 4.2.0. https://CRAN.R-project.org/ package=ggthemes

CDC. (n.d.). Measles cases and outbreaks. https://www.cdc.gov/measles /cases-outbreaks.html

Chatlani, S. (2020). Explainer: What is "flattening the curve"? And why are we "social distancing"? *KPBS*, March 18. https://www.kpbs.org /news/2020/mar/18/explainer-what-flattening-curve-and-why-are -we-soc/

Chetty, R., Hendren, N., Jones, M. R., & Porter, S. R. (2020). Race and economic opportunity in the United States: An intergenerational perspective. *Quarterly Journal of Economics, 135*(2), 711–783.

Conley, D. (2010). *Being Black, living in the Red: Race, wealth, and social policy in America*. University of California Press.

Cunningham, B. (2003). Re-thinking objectivity. *Columbia Journalism Review*. https://archives.cjr.org/feature/rethinking_objectivity .php

Dennis, M. L., Chan, Y., & Funk, R. R. (2006). Development and validation of the GAIN short screener (GSS) for internalizing, externalizing and substance use disorders and crime/violence problems among adolescents and adults. *American Journal on Addictions, 15*, 80–91. https://doi.org/10.1080/10550490601006055

Dragulescu, A., & Arendt, C. (2020). xlsx: Read, write, format excel 2007 and excel 97/2000/XP/2003 files. R package version 0.6.3. https:// CRAN.R-project.org/package=xlsx

Eggerston, L. (2010). Lancet retracts 12-year-old article linking autism to MMR vaccines. *Canadian Medical Association Journal, 182*(4), E199–E200.

Enberger, D. (2017). Is science broken? Or is it self-correcting? *Slate*, August 21. https://slate.com/technology/2017/08/science-is-not-self-correcting-science-is-broken.html

Entman, R. (1993). Framing: Toward clarification of a fractured paradigm. *Journal of Communication, 43*(4), 51–58.

Geiger, A. W. (2019). Key findings about the online news landscape in America. *FactTank, Pew Research Center*, September 11. https://www.pewresearch.org/fact-tank/2019/09/11/key-findings-about-the-online-news-landscape-in-america/.

Goffman, E. (1974). *Frame analysis: An essay on the organization of experience*. Harvard University Press.

Harford, T. (2021). *The data detective: Ten easy rules to make sense of statistics*. Riverhead Books.

Hedley, A. (2018). Data visualization and population politics in *Pearson's Magazine*, 1896–1902. *Journal of Victorian Culture, 23*(3), 421–441.

Lai, K. K. R., & Collins, K. (2020). Which country has flattened the curve for the coronavirus? *The New York Times*, March 19. https://www.nytimes.com/interactive/2020/03/19/world/coronavirus-flatten-the-curve-countries.html

Marcus, A., & Oransky, I. (2020). Science journals are purging racist, sexist work. Finally. *Wired*, September 17. https://www.wired.com/story/science-journals-are-purging-racist-sexist-work-finally/

Marrelli, M., & Zimmer, K. (2018). Twitter's fake news epidemic isn't caused by bots. It's caused by you. *Vice News*, March 8. https://www.vice.com/en/article/bj5ebz/twitters-fake-news-epidemic-isnt-caused-by-bots-its-caused-by-you.

Mervosh, S. (2020). Gov. Mike DeWine of Ohio tests positive, then negative, for coronavirus. *New York Times*, August 7. https://www.nytimes.com/2020/08/06/us/mike-dewine-coronavirus.html?campaign_id=2&emc=edit_th_20200807&instance_id=21064&nl=todaysheadlines&regi_id=69604087&segment_id=35491&user_id=96153b730df88cfb6f14172c669a362e

Moore, D. S., & McCabe, G. P. (1993). *Introduction to the practice of statistics* (2nd ed.). Freeman.

National Academies of Sciences, Engineering, and Medicine. (2019). *Reproducibility and replicability in Science*. The National Academies Press. https://doi.org/10.17226/25303

Nuzzo, R. (2018). Tips for communicating statistical significance. https://www.nih.gov/about-nih/what-we-do/science-health-public-trust/perspectives/science-health-public-trust/tips-communicating-statistical-significance

Owens, D. C., & Fett, S. M. (2019). Black maternal and infant health: Historical legacies of slavery. *American Journal of Public Health*, *109*(10), 1342–1345.

Plesser, J. E. (2018). Reproducibility vs. replicability: A brief history of a confused terminology. *Frontiers in Neuroinformatics*, *11*, 1–3. https://doi.org/10.3389/fninf.2017.00076

Psychological Reports. (2020). Retraction notice. https://doi.org/10.1177/0033294120982774

Puttnam, D. (2013). Does the media have a "duty of care"? *TedXHousesOfParliament*, June. https://www.ted.com/talks/david_puttnam_does_the_media_have_a_duty_of_care?language=en

R Core Team. (2019). *R: A language and environment for statistical computing*. R Foundation for Statistical Computing. https://www.R-project.org/

Roe, Dan. (2017). Inside the Nike Vaporfly: the shoe so good it almost got banned. Runner's World, January 31. Available at: https://www.runnersworld.com/gear/a30730107/nike-vaporfly-tech/.

RStudio Team. (2018). *RStudio: Integrated development for R*. RStudio, Inc. http://www.rstudio.com/

Rykiel, E. J. (2001). Scientific objectivity, value systems, and policymaking. *BioScience*, *51*(6), 433–436.

Saini, A. (2019). The disturbing return of scientific racism. *Wired*, June 12. https://www.wired.co.uk/article/superior-the-return-of-race-science-angela-saini

Society of Professional Journalists. (n.d.). Code of ethics. https://www.spj.org/ethicscode.asp

Soetaert, K., Petzoldt, T., & Setzer, R. W. (2010). Solving differential equations in R: Package deSolve. *Journal of Statistical Software*, *33*(9), 1–25. http://www.jstatsoft.org/v33/i09/; https://doi.org/10.18637/jss.v033.i09

Taylor, J. K. (2020). Structural racism and maternal health among Black women. *Journal of Law, Medicine and Ethics*, *48*(3), 506–517.

The COVID Tracking Project. (n.d.). The COVID racial data tracker. https://covidtracking.com/race

Wallace, M., Crear-Perry, J., Richardson, L., Tarver, M., & Theall, K. (2017). Separate and unequal: Structural racism and infant mortality in the US. *Health and Place*, *45*, 140–144.

Wickham, H. (2016). *ggplot2: Elegant graphics for data analysis.* Springer-Verlag New York.

Wickham, H. (2019). stringr: Simple, consistent wrappers for common string operations. R package version 1.4.0. https://CRAN.R-project.org/package=stringr

Wickham, H., François, R., Henry, L., & Müller, K. (2020). dplyr: A grammar of data manipulation. R package version 0.8.5. https://CRAN.R-project.org/package=dplyr

Wickham, H., François, R., Henry, L., & Müller, K. (2020). dplyr: A grammar of data manipulation. R package version 0.8.5. https://CRAN.R-project.org/package=dplyr

Wickham, H., & Henry, L. (2020). tidyr: Tidy messy data. R package version 1.0.2. https://CRAN.R-project.org/package=tidyr

WKRC. (2020). DeWine tests positive for COVID-19 ahead of Trump visit; second test comes back negative. *WKRC*, August 6. https://local12.com/news/local/gov-dewine-tests-positive-for-covid-19-ahead-of-trump-visit-cincinnati

Zajechowski, M. (n.d.). The eight values that will make your content "news worth". *Digital Third Coast*. https://www.digitalthirdcoast.com/blog/values-content-newsworthy

Index

AAAS, *see* American Association for the Advancement of Science

Absolute risk, 35

Accuracy, 126–127

Albert, Jim, 73

American Association for the Advancement of Science (AAAS), 143

American Time Use Survey (ATUS), 79–82

Annany, Mike, 24

Application programming interfaces (APIs), 5

Aschwanden, Christie, 52

ATUS, *see* American Time Use Survey

Bain, Marc, 56

Binge-watching addiction (television-viewing habits), 75–90

  American Time Use Survey on, 80–82

  claim for, 77

  coverage considerations, 87–90

  evidence, 80

  experts view on, 84

  health impacts associated with, 78–79

  measure of impact, 79

  Netflix survey on, 82–83

  prior belief on, 86

  reflect on our own viewing habits, 87

  review for, 89–90

  stats + stories podcasts, 90

  story, 77

  survey on, 84–86

Bornn, Luke, 73

Both-siderism, 164

Bugeja, Michael, 160–161

Cairo, Alberto, 121

Carrying capacity of system, 19

Cascade, 95–97

Case–control, 37

Causality, 30

Chow, Denise, 12

Climate change, 164

Cohort, 37

Conclusions from data, 6–7

Covariates, 37

COVID-19 spreads in community, 107–121, 165–166

  assumptions, 112

  claims on, 109–110

  coda, 119–120

  coverage considerations, 116–119

  flatten the curve concept and, 108–121

  forced quarantine and, 110

  measure of impact, 111

  prior belief on, 115

quality/strength of evidences on, 113–115
reasons for claims on, 110–111
reported evidences on, 112–113
review/recap, 119
R value in, 109–110
simulation and, 110–115
social distancing and, 110–119
stats + stories podcasts, 120–121
Stevens' on, 110–121
story, 108–109
support for claims on, 111–115
COVID testing in United States, 123–137
claims on, 125–126
community with higher rate of infection, 130–131
community with low rate of infection, 129–130
coverage considerations, 132–135
impact, 132
measure of impact, 126–127
Ohio Governor Mike DeWine, 123–137
prior belief on, 129
quality/strength of evidence on, 129
rapid, less accurate test, 129–131
reported evidence on, 128
screening errors, 127
screening for, 127–129
slower, more accurate test, 130, 131
stats + stories podcasts, 135–137
story on, 124–125

Data
describing, 5–6
generation/producing, 4–5
relevance of, 6–7
Data & Society research group, 104
Department of Economic and Social Affairs (ECOSOC), 14
Describing data, 5–6

DeWine, Mike, 123–137
Distributions, 5–6
Donovan, Joan, 104

Entman, Robert, 68
Extrapolation, 15–16, 48–49

Fake news on social media, tracking spread of, 91–106
claim for, 93–94
coverage considerations, 101–105
measure of impact, 95–97
MIT Media Lab researchers on, 94
previous research on, 100–101
prior belief on, 100
quality/strength of evidence on, 98–100
reflect on our own views, 101
reported evidence on, 97–98
research on, 94
review for, 105
stats + stories podcasts, 105–106
story, 92–93
False negative error, 65
False positive error, 65
Fett, Sharla, 167
Fisher, Nick, 136
Flatten the curve, 108–121
Flowers, Andrew, 135–136
Flynn, James, 162
Forced quarantine, 110
Freeman, Alexandra, 135

Generalizability, 157
Georgiou, Andreas, 24
Gigerenzer, Gerd, 136
Global warming, 163

Halderen, Gemma Van, 24
Hammonds, Evelyn, 162
Hansen, Mark, 24, 106
Harford, Tim, 167
Harkness, Timandra, 90

Harris, Richard, 129
Hedley, Alison, 21, 120–121
Hulu, 78–79
Husted, Jon, 124

Inputs, 17
Interpolation, 15–16
Interval estimate, 44

Journalism, 7–9, 164–168
Journalistic boosterism, 69
Journalists, 2, 7–9, 160–168

Kipchoge, Eliud, 71
Kovach, Bill, 2
KPBS, 119

Lavange, Lisa, 25
Level of significance, 65
Longitudinal studies, 37
Ludwin, Rick, 90
Lurking variables, 47
Lynn, Richard, 161–162

Makulec, Amanda, 120
Mayer, Brian, 25
Mean, 40
Median, 17, 40
Menczer, Filippo, 103, 105
Mervosh, Sarah, 124
MMR vaccine, 164–165
Mode, 40
Mulrow, Jeri, 25
Multinomial logistic regression, 43
Myrick, Jessica Gall, 52, 90

Nagaraja, Chaitra, 24
Netflix, 77, 82–90
News, 95
Nussbaum, Barry, 25

Objectivity, 161
Observational studies, 36–37

Observations, 5
Odds, 34
Odds ratio, 34–35
Open Science Collaboration
    (OSC), 143
OSC, *see* Open Science Collaboration
Outputs, 17
Owens, Deidre Cooper, 167

Page, Danielle, 77
Parameter, 36
Pew Research Center, 103
Point estimate, 44
Polymerase chain reaction, 133
Population, 35–36
Populations growth, 11–25
    behavior as consequence, 20–21
    coverage on, 21–24
    estimation, 12–13
    evidence, 17–18
    exponential *vs.* logistic, 19
    headline, 13–14
    impact, 15
    model, 17–18
    prediction, 15–18
    projections, 14–15, 19
    quality/strength of evidence, 18
    reasonable/unsurprising, 18–19
    result matter, 19–21
    review of news story on, 24
    source of news, 14–15
    statistics, analysis of, 15
    stats + stories podcasts, 24–25
Prediction interval, 17
Probability, 34
Pullinger, John, 24
P value, 65, 147

Quarantine, 110

Racism, 167
Relative risk, 35
Reliability, 32

Repeatability, 156
Replicabilibity, 156
Reproducibility in psychology
    studies, 139–157
    change in behavior, 150–151
    claims on, 142
    coverage considerations, 151–154
    matter, 150–151
    measure of impact, 144
    Open Science Collaboration
        on, 143
    population *vs.* individual
        perspectives, 150
    quality/strength of evidence for,
        147–149
    reasonability of claims on, 149
    reasons for claims on, 143–144
    reported evidence for, 144–147
    review, 154–156
    stats + stories podcasts, 157
    story, 140–142
Reverse causality, 30
Risk, 6
Rosenstiel, Tom, 2
Rumor, 95, 98–106
Running shoe by Nike, 55–73
    control in, 62–63
    design of, 60–65
    evaluation, 67–68
    evidence, 60–63
    experiment on, 58–65
    headline for story, 57
    impact, measure of, 60
    news coverage, 68–71
    prior belief on technology
        associated with, 66
    quality/strength of evidence,
        63–65
    randomization in, 61–62
    replication in, 62
    researchers at University of
        Colorado on, 57–58
    research on, 57–73
    review of story, 71–72

stats + stories podcasts, 73
    testing process for, 64–65
Rushton, John Philippe, 161–162
R value, 109–110
Ryan, Louise, 135

Saini, Angela, 161
Sample, 35–36
Santos, Rob, 24
Scales of measurement, 32–33
Schneider, Claudia, 135
Schwarz, Alan, 73
Schweinfest, Stefan, 24
Searling, Linda, 28
Social distancing, 110–119
Social media and mental health,
    27–53
    adjusted analysis, 44
    adverse outcome, 29–30
    causality, 30
    data collection, 31, 37–38
    designs, observational studies,
        37–45
    evidence, 37–42
    extrapolation, 48–49
    impact, 31–35
    longitudinal studies on, 37
    mental health problems, baseline
        rates of, 46
    news coverage, 49–52
    observational studies on,
        36–45
    odds/odds ratios, 34–35
    odds of problems, increase in,
        45–46
    population, survey on, 35–37
    prior belief about, 46
    quality/strength of evidence,
        43–45
    reverse causality, 30
    risk, 34
    sample, 35–37
    statistics on, 35–45
    stats + stories podcasts, 52–53

survey of US adolescents on, 52
unadjusted analysis, 44
variables, 31–34
youths with, 45–48
Standard deviation(s), 42
Stevens, Harry, 108–121

Tashiro, Ty, 52, 90
Television watching, 77–90
Thompson, John, 24
Trewin, Dennis, 136
Tucker, Joshua, 106
Twitter, 92–106

Uncertainty, 6–7

United Nations Statistical
Commission, 14
UN report, 23

Validity, 32
Variables, 5–6
Vosoughi, Soroush, 104

Wang, Hansi Lo, 24
Wasserstein, Ron, 25
Weight, 36
Wilson, Steven Lloyd, 105–106
Wright, Tommy, 24

Young, Linda, 25

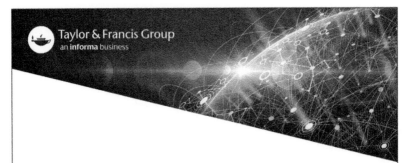

Printed and bound by CPI Group (UK) Ltd, Croydon, CR0 4YY

01/11/2024

01782620-0001